The Ground Sloth
Megalonyx

TRANSACTIONS
of the
AMERICAN PHILOSOPHICAL SOCIETY
Held at Philadelphia
For Promoting Useful Knowledge
Volume 100, Part 4

The Ground Sloth
Megalonyx

(Xenarthra: Megalonychidae) from
the Pleistocene (Late Irvingtonian)
Camelot Local Fauna, Dorchester County,
South Carolina

Steven E. Fields

AMERICAN PHILOSOPHICAL SOCIETY
Philadelphia • 2010

ISBN: 978-1-60618-004-4

US ISSN: 0065-9746

Library of Congress Cataloging-in-Publication Data

Fields, Steven E., 1966-
  The ground sloth Megalonyx (xenarthra: megalonychidae) from the Pleistocene (late Irvingtonian) Camelot local fauna, Dorchester County, South Carolina / Steven E. Fields.
     p. cm. -- (Transactions of the American Philosophical Society held at Philadelphia for promoting useful knowledge; v. 100, pt. 4)
  Includes bibliographical references and index.
  ISBN 978-1-60618-004-4
  1. Megalonychidae, Fossil--South Carolina--Dorchester County. 2. Xenarthra, Fossil--South Carolina--Dorchester County. 3. Paleontology--Pleistocene. I. Title.
  QE882.E2F54 2010
  569'.31--dc22
                                        2010027220

Table of Contents

Preface

The late Irvingtonian (400–450 ka) Camelot site in the Coastal Plain of South Carolina has yielded numerous taxa of fossil mammals. Among these are over 250 teeth and postcranial elements of juvenile and adult specimens of the ground sloth *Megalonyx*, representing one of the largest samples of *Megalonyx* in the southeastern United States. Morphometric data are presented in detail and they suggest the population was in a state of transition towards the larger size that characterized the later Rancholabrean forms. Quantitative characters show considerable overlap, and qualitative characters alone may not be adequate to distinguish species. The validity of currently recognized species of *Megalonyx* from the Pleistocene needs to be reconsidered.

The Ground Sloth
Megalonyx

Introduction

In the late 1790s Thomas Jefferson presented a paper to the American Philosophical Society on skeletal remains of an animal from western Virginia. He described the bones from an animal "of the clawed kind."(Jefferson 1799). A large claw core in the remains prompted Jefferson to name the beast *Megalonyx*, meaning "great claw." Jefferson assumed that the creature must have been some type of large carnivore, such as a lion. However, similar finds in South America led to the correct identification of the animal as a large ground sloth. The species was subsequently named *Megalonyx jeffersonii* (Desmarest 1822) in honor of Thomas Jefferson. Since that time, numerous other species of *Megalonyx* have been described and synonomized.

Fossil ground sloths of the genus *Megalonyx* range from the late Miocene (Hemphillian, ca. six million years ago) to the late Pleistocene (Rancholabrean, ca. 12,000 years ago) (McDonald et. al 2000, Woodburne 2004). The genus is known from over 152 sites in North America (McDonald 2003) from Florida (Hulbert 2001) to Alaska (Stock 1942). Fossils of *Megalonyx* from the late Blancan, Irvingtonian, and Rancholabrean North American Land Mammal Ages (NALMA) have been collected in South Carolina. Up to three species of *Megalonyx* from the Plio-Pleistocene of South Carolina may be represented by the chronospecies *M. leptostomus, M. wheatleyi, M. jeffersonii* as proposed by McDonald (1977).

Although *Megalonyx* is a common component of many North American Pleistocene faunas, fossil elements of ground sloth are usually not this abundant (H.G. McDonald, pers. comm.). Typically, isolated teeth comprise the representative *Megalonyx* fossils for any given site. Without associated mandibles, the molariform teeth of sloths are difficult to assign position within the ramus. As such, isolated teeth are usually of little value besides documenting occurrence of the taxon at a particular locality.

More than 200 years after Thomas Jefferson made his discovery, a new site in the coastal plain of South Carolina has yielded an unusually large sample of *Megalonyx* remains. This paper reports over 250 elements of *Megalonyx* from the late Irvingtonian Camelot Local Fauna in Dorchester County, South Carolina. Initial study suggested that this sample was a population of smaller-sized *M. jeffersonii*, but other characters are consistent with *M. wheatleyi*. It appears that *Megalonyx* from Camelot represent a transitional form within the chronospecies *M. wheatleyi* → *M. jeffersonii*. Assignment to species is therefore difficult. Comparisons to other Plio-Pleistocene *Megalonyx* specimens from various North American sites reveal significant overlap in measurement values. Assessment of most qualitative characters is subjective. Thus, a review of the current taxonomy of Pleistocene *Megalonyx* is the subject of another paper that will partially compose the author's Ph.D. dissertation. The present paper provides the comprehensive measurements of the Camelot sample for descriptive purposes and compares them to other *Megalonyx* to demonstrate the transitional nature of the Camelot population.

Study Site

The Camelot Local Fauna is a late-middle Pleistocene (late Irvingtonian) site near Harleyville in Dorchester County, South Carolina (Figure 1). Fossil material was collected from a channel-fill in the unnamed sands and clays that overlie the Tupelo Bay Formation (Geisler et al. 2005) in the Giant Cement Plant quarry about 4 km north of Harleyville. The age of the site (400–450 ka) was determined biostratigraphically by several of the taxa thus far identified from the fauna.

Camelot is temporally equivalent to the Coleman 2A Local Fauna in Florida (Kohn et al. 2005). Ironically, no sloth fossils were recovered from Coleman 2A (Martin 1974). No specimens of *Bison*, a Rancholabrean indicator species, have been found at Camelot, but the fauna includes *Canis armbrusteri*, which became extinct at the end of the Irvingtonian, and rabbits of the genus *Lepus*, which disappeared from the southeastern United States at the end of the Irvingtonian (Kurtén and Anderson 1980). The only ground sloth thus far found at Camelot is *Megalonyx*. A sabercat from the Camelot Local Fauna is identified as *Smilodon* cf. *S. fatalis*, intermediate in size between the later *S. fatalis* from Rancho La Brea and the earlier *S. gracilis* from the Blancan of Florida. Additionally, the *Smilodon* from Camelot differs from *S. gracilis* in that it has a single-rooted lower p3, which is double-rooted in *S. gracilis* and is lost in the Rancholabrean samples. Therefore, the dentition of *Smilodon* from Camelot represents an intermediate condition. In this paper I present morphometric data that suggest that the *Megalonyx* specimens from Camelot also represent an intermediate form in the series of chrono-species currently recognized.

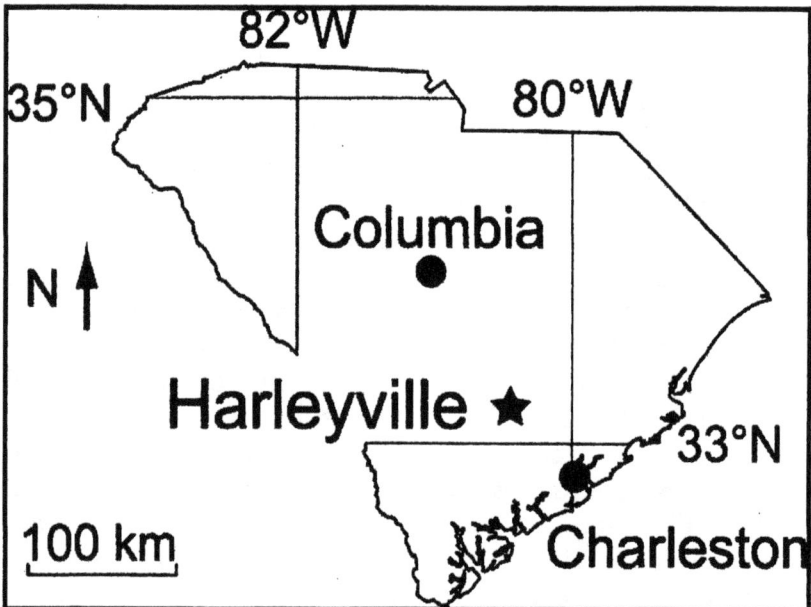

FIGURE 1. Location of the Camelot Local Fauna. The late Irvingtonian site is near the city of Harleyville on the Coastal Plain of South Carolina. Map from Kohn et al. (2005).

Materials and Methods

All *Megalonyx* fossils from Camelot are housed at the South Carolina State Museum (SCSM) in Columbia, South Carolina, and are cataloged under accession numbers SCSM 2003.75 and SCSM 2004.1. I measured all catalogued *Megalonyx* fossils with Mitutoyo MyCAL 0–8"/0200mm digital calipers to the nearest 0.1 mm. I measured specimens exceeding 200 mm in length with a "measuring box" as described by von den Driesch (1976 p.10) to the nearest 0.5 mm. To facilitate comparisons with other specimens I present these measurements as in McDonald (1977), McDonald et al. (2000), and McDonald et al. (2001), and I included standard univariate (summary) statistics where appropriate. Unless otherwise stated, all measurements are in millimeters (mm). The data are compared to *Megalonyx* measurements from other localities to demonstrate the transitional nature of the late Irvingtonian Camelot population.

Institutional acronyms are as follows: **AM** and **AMNH**, American Museum of Natural History; **ANSP**, Academy of Natural Sciences at Philadelphia; **BYU**, Brigham Young University; **ChM**, Charleston Museum; **CMN**, Canadian Museum of Nature; **DMNH**, Dayton Museum of Natural History; **F:AM**, American Museum of Natural History; **IMNH**, Idaho Museum of Natural History; **ISUM**, Idaho State University Museum; **LACM**, Los Angeles County Museum of Natural History; **MR**, Museum of the Rockies; **NMC**, National Museum of Canada; **NWSM**, North Central Washington

Museum; **PPHM**, Panhandle Plains historical Museum; **SCSM,** South Carolina State Museum; SMU, Schuler Museum, Southern Methodist University; **SUI**, State University of Iowa; **TMM**, Texas Memorial Museum; **UCMP**, University of California Museum of Paleontology; **UF**, University of Florida Museum of Natural History; **UMMP**, University of Michigan Museum of Paleontology; **UN**, University of Nebraska State Museum; **USNM**, Smithsonian National Museum of Natural History; **UW**, Burke Museum, University of Washington. Other abbreviations are: **L.F.**, Local Fauna; **SD,** standard deviation; **n**, sample size; **~**, approximate measurement; **+**, part of specimen is broken off and/or missing.

Results and Discussion

Cataloged fossils of *Megalonyx* from the Camelot site totaled approximately 253 skeletal elements including cranial fragments, mandibles, teeth, vertebrae, ribs, scapulae, humeri, ulnae, radii, carpals, metacarpals, femora, fibulae, tibiae, tarsals, metatarsals, and phalanges of the manus and pes. The sample has both juveniles (approximately 48 bones with epiphyses not fused) and adults. The minimum number of individuals (MNI) was six, based on available elements.

Cranium, mandibles, and dentition—No complete crania have been recovered from the Camelot site. Two sets of cranial fragments are cataloged under 2003.75.425 and 2004.1.30. The latter retains an upper right dentition of the caniniform and four molariforms. The remaining portion of the skull is in fragments and not measurable. Two complete mandibles and two partial mandibles are represented (see Figure 2). Table 1 provides measurements for Camelot mandibles compared to mandibles from other sites.

McDonald (1977) used ratios of mandible measurements to assess sexual dimorphism in *Megalonyx*. Specifically, he used the ratio of diastema length to alveolar length of molariform row and the ratio of caniniform length to diastema length. According to McDonald (1977) the ratio of caniniform length to diastema length ≥ 1.00 in male *Megalonyx* and < 1.00 in females. Also, the ratio of diastema length to alveolar length of cheek teeth is generally < 5.50 in males and > 6.50 in females. Ratios were calculated for both complete Camelot

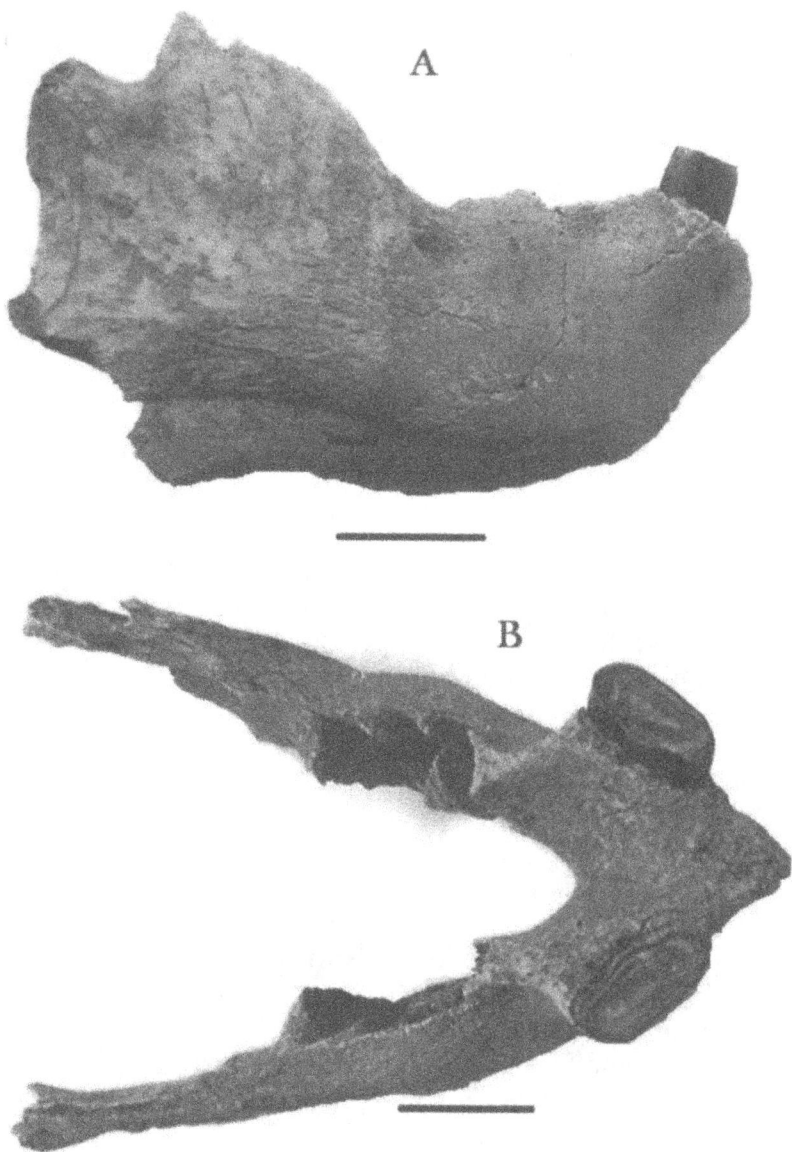

FIGURE 2. Mandible of *Megalonyx* from Camelot L.F. (SCSM 2004.1.26).
A, right lateral view, **B**, occlusal view. Scale bar = 5cm.

TABLE 1. Measurements of *Megalonyx* mandibles from Camelot compared to specimens from other localities (data from McDonald 1977): 1. Length of mandible from anterior edge of caniniform to posterior edge of angular process; 2. Depth of mandible below first molariform; 3. Length of diastema from posterior edge of caniniform to anterior edge of first molariform; 4. Width of diastema; 5. Alveolar length of cheek teeth; 6. Length from anterior edge of the caniniform to the posterior edge of the third molariform. Summary statistics are included for Camelot samples.

Megalonyx from Camelot L.F. (SCSM)		1	2	3	4	5	6
2003.75.424		X	71.3	X	X	64.6	X
2003.75.330		X	68.9	X	X	69.7	X
2004.1.3.1		234.5	72.8	36.8	8.6	63.9	121.5
2004.1.26.1		220.0	74.5	44.7	8.7	65.6	132.5
Camelot L.F. Summary Statistics	Mean	227.3	71.9	40.7	8.6	66.0	127.0
	SD	10.3	2.4	5.6	0.1	2.6	7.8
	n	2	4	2	2	4	2
Megalonyx leptostomus							
USNM 12669	Hagerman, Idaho	178+	59.0	24.3	14.4	51.6	102.5
USNM 13477	Hagerman, Idaho	168+	65.4	30.2	12.2	~54	111.5
UF 223806	Haile 7G, Florida	225.0	63.0	36.5	9.3	52.5	106.0
Megalonyx wheatleyi							
USNM 11633	Port Kennedy Cave, Pennsylvania	206+	76.8	41.8	12.6	67.8	130.0
ANSP 180	Port Kennedy Cave, Pennsylvania	X	77.8	37.5	10.2	63.8	137.3
F:AM 99187	McLeod Limerock Mine, Florida	251.2	75.0	37.9	11.2	61.9	127.5
F:AM 99192	McLeod Limerock Mine, Florida	~220	~67.5	29.5	9.5	62.2	~117
Megalonyx jeffersonii							
ISUM 15151	American Falls, Idaho	272.0	82.7	37.3	10.3	70.2	138.4
ISUM 23034	American Falls, Idaho	298.5	93.2	45.1	10.8	69.5	145.5
UCMP 41-4-53	Deer Creek, OK	262.2	72.4	39.6	8.6	65.7	143.3
UCMP 21429	Rancho La Brea, California	256.0	85.7	51.0	7.9	67.5	138.9
UF 16383	Waccassasa River, Florida	265+	90.0	34.9	13.9	72.3	142.1

11

mandibles (SCSM 2004.1.3 and 2004.1.26) and resulting values (see Table 2) suggested they were both females (see also
McDonald 1977:62–64). There is further variation noted in
both complete Camelot *Megalonyx* mandibles, as SCSM
2004.1.26 is wider (posterior ends of mandibles are farther
apart) than SCSM 2004.1.3.

Some of the mandibular elements provided an opportunity
to assess hypsodonty in the Camelot population. Bargo et
al. (2006) considered hypsodonty (crown height of teeth) in
Pleistocene ground sloths as an indicator of dietary prefer-
ences, habitat, and habits. Increased hypsodonty in sloths
may also be related to the absence of enamel in xenarthran
teeth. Bargo et al. (2006) considered increased hypsodonty in
some mylodontid ground sloths to possibly be the result of
ingesting abrasive soil as they excavated and ate roots and
tubers. Increased hypsodonty in *Paramylodon harlani* may
have been due to a change over time from forested to open
habitats (see McDonald 1995). Differences in hypsodonty
between *Megatherium* and *Eremotherium* were likely due to
differences in geographic distribution. However, while Bargo
et al. (2006) included samples from mylodontids and mega-
theriids, they did not include *Megalonyx* or any other megalonychids, citing inaccessibility to specimens. However, to
supplement the work of Bargo et al. (2006), Fields (2009) pres-
ents hypsodonty indices for three species of Plio-Pleistocene
megalonychids.

Bargo et al. (2006) calculated hypsodonty indices using
the following formula: HI = DM/LTR where HI is the hypsodonty index, DM is the depth of the mandible, usually
measured below the third molariform, and LTR is the length
of the molariform tooth row. Hypsodonty indices (HI) for the
Camelot specimens are listed in Table 3. The mean HI for the
Camelot sample was 1.02 (SD = 0.06). This value is identical to
the mean HI for *Megatherium americanum* reported in Bargo
et al. (2006). While the South American *Megatherium* is con-
siderably larger than *Megalonyx*, both are typically regarded
as browsers, or at least not as grazers on coarse, gritty vegeta-
tion (McDonald 1977, Kurtén and Anderson 1980, McDonald
1995). However, specific factors that directly influence hypsodonty are difficult to identify (Bargo et al. 2006). Therefore,
hypsodonty indices alone may not accurately reflect dietary
habits in *Megalonyx* or other ground sloths.

TABLE 2. Mandibular measurements and ratios of *Megalonyx* from Camelot L.F. to assess sexual dimorphism. Abbreviations: DL = diastema length; ALM = alveolar length of molariforms; CL = caniniform length.

SCSM No.	Element	DL	ALM	CL	DL/ALM	CL/DL	Sex
2003.75.424	ramus, partial	X	64.6	31.2	Unavailable	Unavailable	Unknown
2003.75.330	ramus, fragment	X	69.7	X	Unavailable	Unavailable	Unknown
2004.1.3.1	mandible	36.8	63.9	30.6	0.575	0.833	Female
2004.1.26.1	mandible	44.7	65.6	32.7	0.682	0.731	Female

TABLE 3. Hypsodonty indices (HI) for *Megalonyx* from Camelot L.F. HI was calculated as depth of mandible (DM) divided by length of molariform tooth row (LTR). For complete mandibles (SCSM 2004.1.3 and SCSM 2004.1.26) HI was calculated for both left and right sides of the mandible, resulting in slightly different values of HI. Measurements are in millimeters (mm).

SCSM No.	Specimen	DM	LTR	HI
2004.1.3	complete mandible (right side)	68.5	63.9	1.07
2004.1.3	complete mandible (left side)	63.7	62.9	1.01
2004.1.26.1	complete mandible (right side)	64.2	64.1	1.00
2004.1.26.1	complete mandible (left side)	62.4	65.6	0.95
2003.75.330	partial mandible	68.9	69.7	0.99
2003.75.424.1	partial mandible	71.3	64.6	1.11
Mean		66.5	65.1	1.02
SD		3.5	2.4	0.06
n		6	6	6

Numerous teeth are represented in the Camelot Local Fauna. Teeth are often common among fossil faunas and assemblages, and such finds are particularly important in the study of mammals. Mammalian dentitions are typically specialized because most reflect adaptations for feeding and other lifestyle habits. However, xenarthran teeth lack enamel and it is uncertain if they are homologous to other mammalian teeth (Bargo et al. 2006). Hence, sloth teeth are labeled as "caniniform" and "molariform." Dentitions of ground sloths vary even among taxonomic families. Normal *Megalonyx* dentition includes two upper caniniforms and eight upper molariforms (one caniniform and four molariforms on each side). The lower jaw includes two caniniforms and six molariforms (one caniniform and three molariforms on each side—see Figure 2). In both the upper and lower jaws the caniniforms and molariforms are separated by a diastema. It is possible to assign isolated caniniforms to position in the skull or mandible, but it is much more difficult to correctly place isolated molariforms. Indeed, variable tooth morphology has prompted the discovery of many isolated teeth being described as new species. Cope (1871) described five species of *Megalonyx* from Port Kennedy Cave in Pennsylvania. That interpretation was later questioned by Hirschfeld and Webb (1968), and Cope's proposed species were ultimately synonomized with *M. wheatleyi* by McDonald (1977).

The Camelot sample of *Megalonyx* teeth comprises at least 13 caniniforms and 27 molariforms (some associated with mandibles). See Table 4 for a summary of teeth and measurements. McDonald (1977) and McDonald et al. (2000) used length and width of the upper and lower caniniforms to distinguish the various taxa within the Megalonychidae. However, scatter diagrams (Figure 3) reveal considerable overlap in caniniform size in *Megalonyx* from various sites. Caniniforms from Camelot *Megalonyx* fall within the range of large *M. wheatleyi and* small *M. jeffersonii.* McDonald (1977) used qualitative characters of caniniforms to distinguish *M. wheatleyi* from *M. jeffersonii.* McDonald (1977: 267) said of *M. wheatleyi,* "...upper caniniform with medial lingual bulge but anterior and posterior longitudinal grooves weak or absent." McDonald (1977:257) said of *M. jeffersonii,* "...upper caniniform with prominent lingual bulge medially located and with well developed anterior and posterior

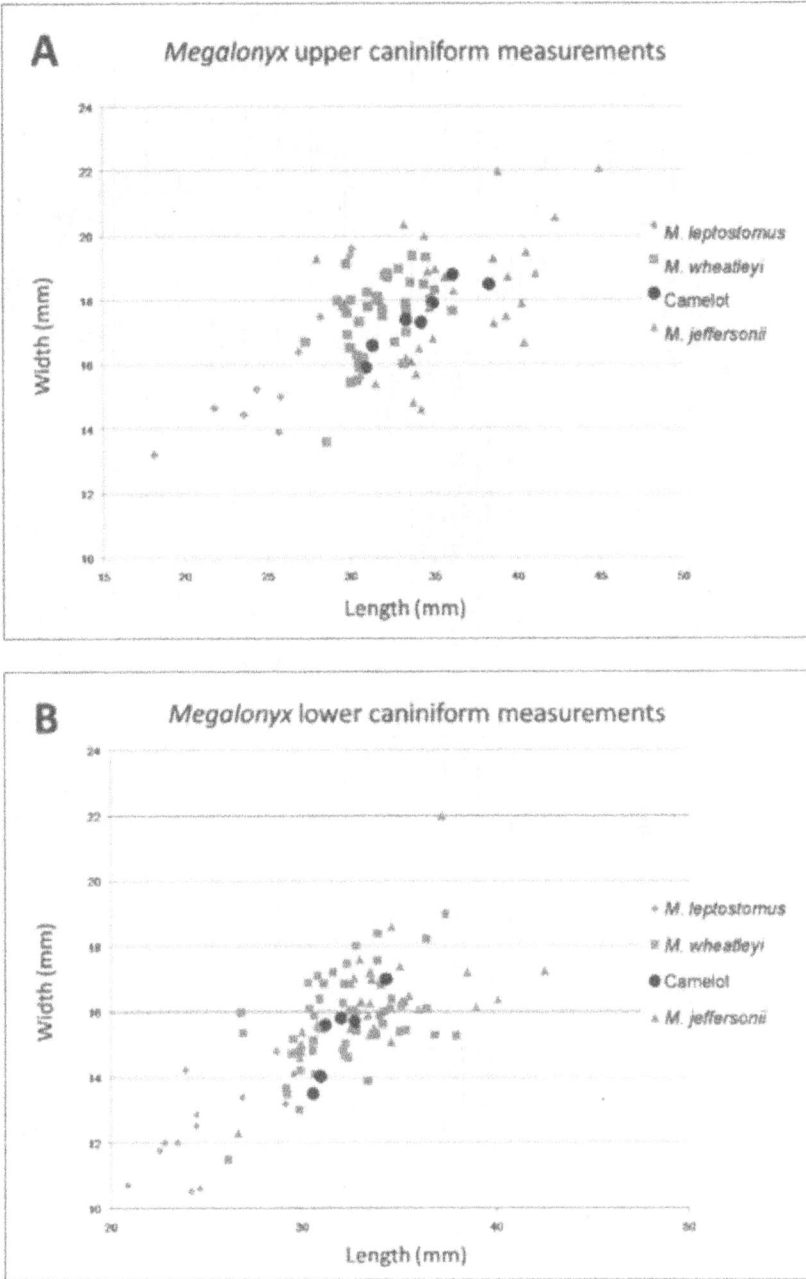

FIGURE 3. Scatter diagram of *Megalonyx* caniniforms. **A**, upper caniniforms. **B**, lower caniniforms). Note that specimens from Camelot fall within the range of both *M. wheatleyi* and *M. jeffersonii*. Plot data for Camelot are listed in Table 4. Plot data for other *Megalonyx* samples are listed in Appendix 1.

TABLE 4. Measurements of *Megalonyx* teeth from Camelot L.F. Abbreviations for caniniforms: UL = upper left, UR = upper right, LL = lower left, LR = lower right; abbreviations for molariforms: M = upper molariform, m = lower molariform. Position, if known, is denoted by numbers 1, 2, 3, or 4. Caniniform measurements are plotted in the scatter diagram (Figure 3).

SCSM No.	Element	Length	Width	Comments
Caniniforms				
2003.75.424	caniniform-LL	31.2	15.6	fragmented ramus associated
2004.1.26	caniniform-LL	32.7	15.7	mandible associated
2004.1.26	caniniform-LR	32.0	15.8	mandible associated
2004.1.3	caniniform-LL	31.0	14.0	mandible associated
2004.1.3	caniniform-LR	30.6	13.5	mandible associated
2004.1.76	caniniform-LL	34.3	17.0	
2003.75.328	caniniform-UL	36.1	18.8	
2003.75.409	caniniform-UL	31.3	16.6	
2004.1.	caniniform-UL	34.9	17.9	
2004.1.129	caniniform-UL	34.2	17.3	
2004.1.77	caniniform-UL	38.3	18.5	
2004.1	caniniform-UR	30.9	15.9	
2004.1	caniniform-UR	33.3	17.4	

Molariforms

2003.75.333	Isolated molariform	16.6	22.8	associated molariforms & jaw fragments
2003.75.333	Isolated molariform	15.2	21.4	associated molariforms & jaw fragments
2003.75.333	Isolated molariform	16.1	23.7	associated molariforms & jaw fragments
2003.75.39	Isolated molariform	12.1	18.7	
2004.1.78	Isolated molariform	15.1	20.8	
2004.1.79	Isolated molariform	16.1	23.4	
2003.75.425.2	Left M2	17.5	23.9	associated with cranial fragments
2003.75.425.2	Right M3	15.4	22.2	associated with cranial fragments
2004.1.3	Left m1	14.6	18.9	mandible associated
2004.1.3	Left m2	14.4	20.2	mandible associated
2004.1.3	Right m1	14.8	19.1	mandible associated
2004.1.3	Right m2	14.8	20.3	mandible associated
2004.1.3	Right m3	15.7	20.5	mandible associated
2004.1.26	Left m3	16.8	23.4	mandible associated
2004.1.26	Right m1	15.3	22.2	mandible associated
2004.1.26	Right m2	15.9	24	mandible associated
2004.1.26	Right m3	17.8	22.4	mandible associated
2004.1.30	Right M1	15.1	19.5	cranium associated
2004.1.30	Right M2	14.5	19.6	cranium associated
2004.1.30	Right M3	15.5	19.7	cranium associated

grooves." Once again, the Camelot sample contains specimens that exhibit characters of each taxon (see Figure 4).

Vertebrae—Fifty-six vertebral elements of *Megalonyx* were cataloged from Camelot. Because elements were isolated it was not possible to assign exact position of each vertebra to the spinal column. However, morphological comparisons allowed assignment of most specimens to general type of vertebra. The sample includes two atlases, four cervical, eight anterior thoracic, 18 posterior thoracic, one partial thoracic, four lumbers, and 15 caudal elements. Four vertebral centra were not assignable to position. See Appendix 2 for a complete list and measurements of all cataloged vertebral elements from Camelot. See also Figure 5.

In all specimens where only one epiphysis of the centrum was fused, the anterior epiphysis was fused and the posterior epiphysis was not. It is not known whether this is a normal in the ontogeny of *Megalonyx*. However, McDonald (1977: p.93) said that epiphyseal fusion of vertebral centra could not be used as an indicator of maturity. Still, at least 30 vertebrae are present in which neither the anterior nor posterior epiphysis is fused.

Ribs—Numerous ribs and rib fragments were recovered from the Camelot site, but as of this writing, only eight sternal rib elements have been prepared and cataloged (six elements under base number SCSM 2003.75: .334, .399, .400, .401, .402, and .403; and two elements under base number SCSM 2004.1: .93 and .94). Additionally, there is a single partial costal rib (SCSM 2004.1.127) and three sternebrae (SCSM 2004.1: .103, .120, .121). McDonald (1977) noted eight pairs of sternal ribs that articulate with the sternebrae in *Megalonyx*, but the Camelot specimens were isolated and in fragments; it was not possible to assign them to position on the body. It is therefore unknown if the sternal rib elements were associated. Selected images of the Camelot rib elements are presented in Figure 6.

Scapula—A single highly fragmented scapula was recovered from the Camelot site. The extremely fragile nature of this specimen prohibited manipulation and positioning for photography. Measurements are provided in Table 5. No measurements for scapular dimensions of *M. wheatleyi* were available, but the size of the Camelot specimen falls between that of *M. leptostomus* and *M. jeffersonii*.

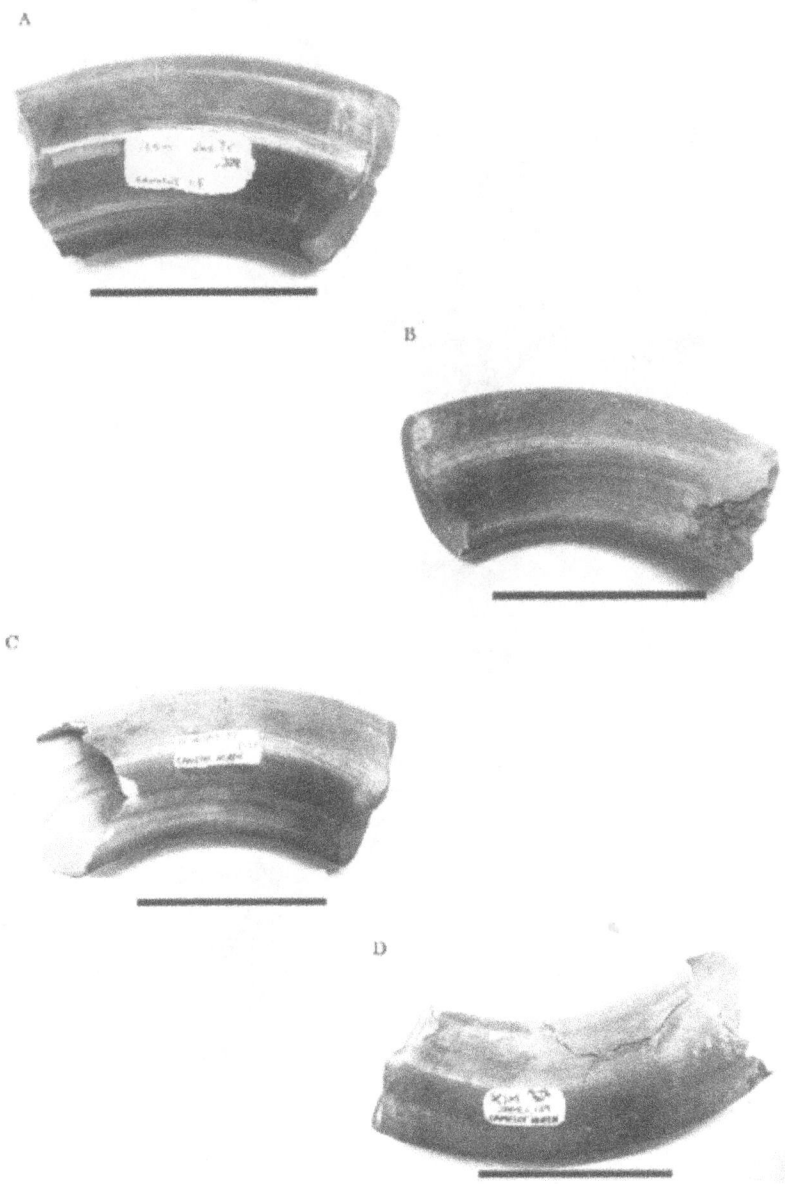

FIGURE 4. Lingual views of upper caniniforms of *Megalonyx* from Camelot L.F. Specimens in this sample show qualitative characters of both *M. wheatleyi* and *M. jeffersonii* as described by McDonald (1977). **A**, SCSM 2003.75.328 displays a prominent medial lingual bulge with well developed anterior and posterior grooves (characters of *M. jeffersonii*). B, SCSM 2004.1.30 has a less prominent medial lingual bulge but weak anterior and posterior grooves (characters of *M. wheatleyi*). **C**, SCSM 2004.1.77 and **D**, SCSM 2004.1.129 appear intermediate in form. Both have a prominent lingual bulge and but weak anterior and posterior grooves. Scale bar in each image = 5 cm.

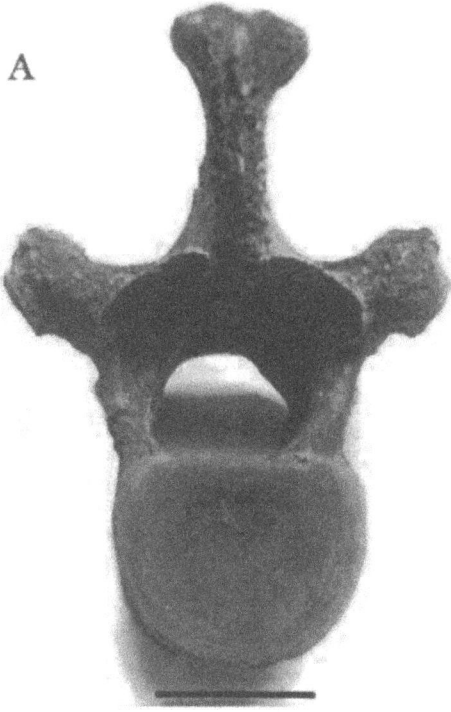

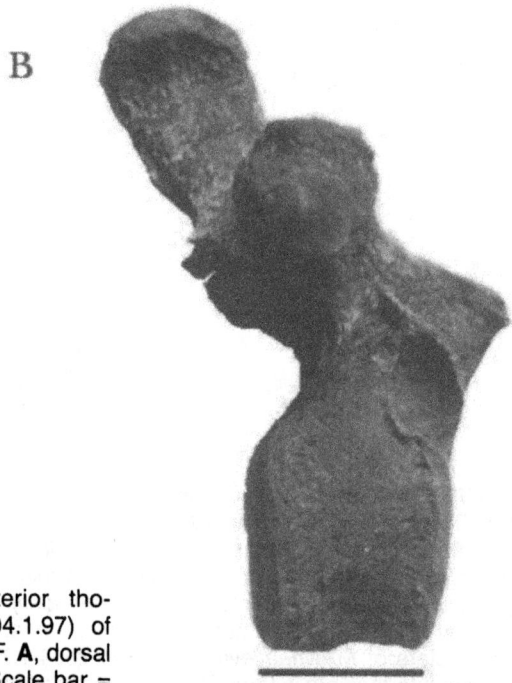

FIGURE 5. Sample posterior thoracic vertebra (SCSM 2004.1.97) of *Megalonyx* from Camelot L.F. **A**, dorsal view. **B**, right lateral view. Scale bar = 5cm.

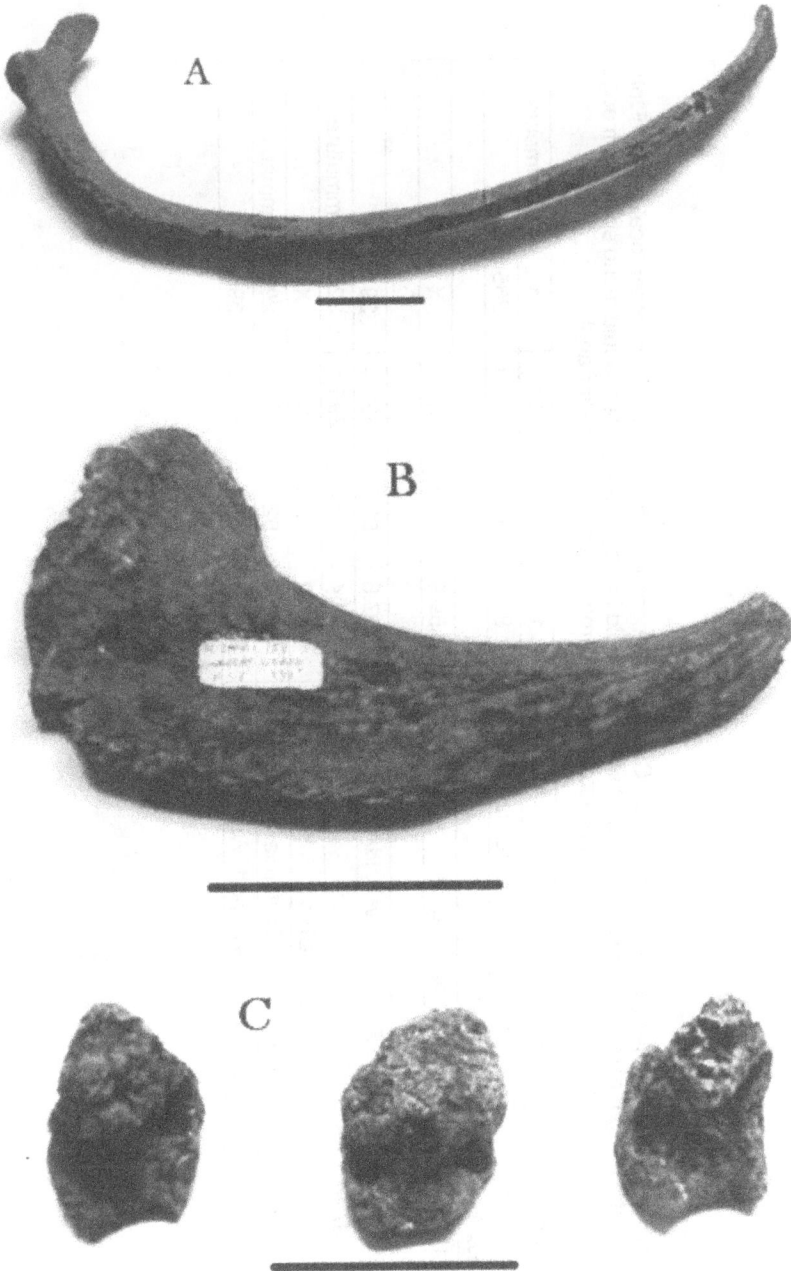

FIGURE 6. Rib elements of *Megalonyx* from Camelot L.F. **A**, SCSM 2004.1.94 rib. **B**, SCSM 2004.1.127 costal rib. **C**, SCSM 2004.1.103, 120, 121 sterne-brae. Scale bar = 5cm.

TABLE 5. Measurements of *Megalonyx* scapula from Camelot L.F. compared to scapulae from other localities (McDonald 1977): 1. Greatest length of glenoid fossa; 2. Greatest width of glenoid fossa; 3. Length of scapula along spine from edge of glenoid to vertebral border; 4. Transverse distance across scapula from anterior to posterior angle.

Catalog number	Locality	1	2	3	4	Comments
SCSM 2003.75	Camelot L.F., South Carolina	83.5	46.8	275.0	265.0	
Megalonyx leptostomus						
USNM 214934	Ringhold, Washington	59.9	37.7	X	X	
Megalonyx jeffersonii						
ISUM 23034	Amer. Falls Reservoir, Idaho	97.6	62.3	323.4	391.0	
ANSP 12497	Big Bone Cave, TN	92.4	56.8	295.4	326.6	immature
DMNH G-25698	Darke County, Ohio	111.0	72.0	X	X	
USNM 10927	Mercer's Cave, California	98.4	52.3	X	290.0	immature
UW 20788	Seattle Tacoma Airport, Washington	91.9	69.1	318.2	395.0	

Clavicle—A single *Megalonyx* clavicle (SCSM 2003.75.391) was collected from the Camelot site (see Figure 7). McDonald (1977) did not include measurements of the clavicle, and no other specimens were available for study. This specimen is 173.0 mm in greatest length 173.0 and 38.5 mm in greatest width. It has the general "boomerang" shape as described by McDonald (1977).

Humerus—No complete humeri were recovered from Camelot, but partial elements and fragments of four specimens afforded some measurements and indication of size. All four specimens had unfused epiphyses, indicating that all were immature individuals (see Figure 8). Although these are immature specimens, their size is comparable to adult specimens of *Megalonyx wheatleyi*, suggesting that adult specimens would be somewhat larger. Such data reinforce the notion that the Camelot population was slightly larger than most *M. wheatleyi* but smaller than most *M. jeffersonii*. Measurements are presented in Table 6. Roth (1990) used dimensions of the humerus to estimate body mass for proboscideans. The same formula is used to estimate body mass in *Megalonyx* (see Table 21 and section on body mass). McDonald (1977) further noted possible sexual dimorphism in *Megalonyx* humeri. Larger diaphysis width was suspected to support a more robust musculature in male animals. Unfortunately, insufficient sample size prohibited the assessment of this trend in the Camelot sample.

Ulna—Five specimens, including one complete (Figure 9) and four partial elements comprised the ulnar sample from Camelot. Dimensions of SCSM 2004.1.84 were somewhat puzzling because the specimen had unfused epiphyses, but some measurements were greater than or equal to adults from the sample (Table 7). This may simply indicate a high degree of size variation within the population. Alternatively, there could be sexual dimorphism with males and females exhibiting significant differences in size. As discussed earlier, sexual dimorphism is measureable in mandibular and humeral elements of *Megalonyx* (McDonald 1977).

Radius—Seven specimens, including two complete adult radii (Figure 10) and six partial juvenile elements are known from Camelot (Table 8). Notably, immature specimens again show similar dimensions to those of adults. It is doubtful that the variation is due to sexual dimorphism because there does

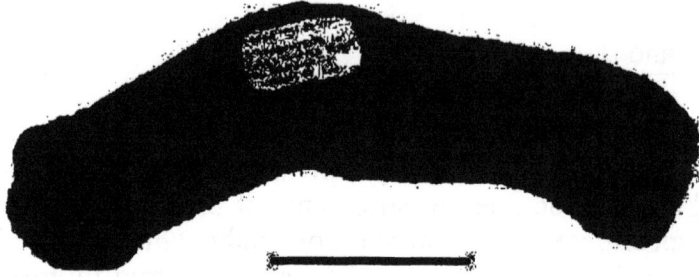

FIGURE 7. *Megalonyx* clavicle (SCSM 2003.75.391) from Camelot L.F. Scale bar = 5cm.

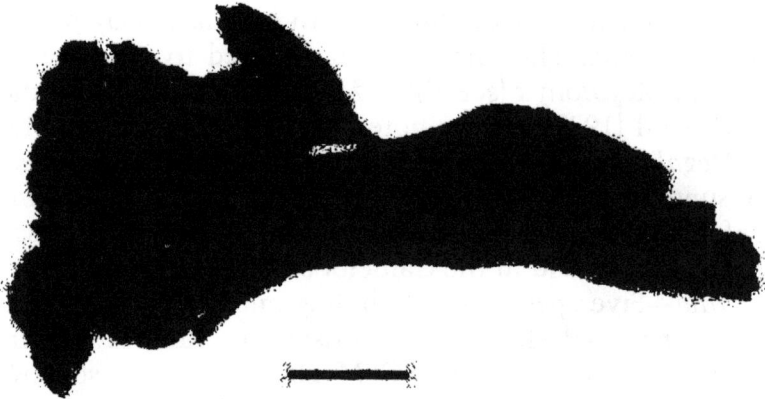

FIGURE 8. Dorsal view of partially reconstructed juvenile *Megalonyx* right humerus (SCSM 2004.1.112) from Camelot L.F. Scale bar = 5cm.

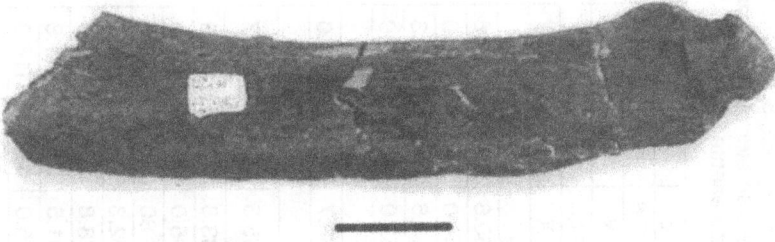

FIGURE 9. Left distal ulna of *Megalonyx* (SCSM 2003.75.387) from Camelot. Scale bar = 5cm.

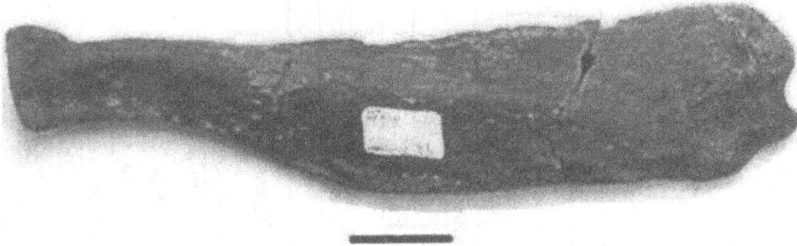

FIGURE 10. Adult left radius of *Megalonyx* (SCSM 2003.75.88) from Camelot L.F. Scale bar = 5cm.

TABLE 6. Measurements of *Megalonyx* humeri from Camelot L.F. compared to specimens from other localities (some data from McDonald 1977): 1. Greatest length between top of head and medial condyle; 2. Greatest transverse diameter of proximal end; 3. Greatest anteroposterior diameter of proximal end; 4. Greatest transverse diameter of head; 5. Length of diaphysis along posterior side; 7. Greatest transverse diameter of diaphysis; 8. Greatest transverse diameter of distal end; 9. Greatest transverse diameter of trochlea. Note: All Camelot humeri had unfused epiphyses i.e. represented juveniles.

Megalonyx from Camelot L.F. (SCSM)		1	2	3	4	5	6	7	8
2003.75.404		X	103.8	81.4	X	307.0	58.6	174+	110.6
2003.75.441		X	110.1	90.5	66.8	X	X	X	X
2004.1.112		X	80.5+	X	X	X	54.4	175+	X
2004.1.124		X	X	X	X	X	X	X	107.7
Megalonyx leptostomus									
USNM 214934	Ringhold, Washington	311.7	83.0	67.6	53.5	249.4	50.6	139.6	80.6
UN 42425	Broadwater, NE	342.0	83.0	61.0	X	X	57.0	145.0	88.0
UF ML-11	Haile 7G, Florida	314.0	76.5	63.0	46.2	271.0	45.6	132.9	79.3
UF ML-8	Haile 7G, Florida	377.0	88.2	76.7	57.9	323.0	59.0	162.6	85.0
Megalonyx wheatleyi									
F:AM 103-1986	McLeod, Florida	420.0	115.9	88.1	61.6	352.6	68.7	X	100.0
Megalonyx jeffersonii									
TMM 30967-1407	Ingleside, Texas	473.1	118.8	102.3	71.8	394.0	85.5	221.3	115.7
ISUM 16462	Amer. Falls Reservoir, Idaho	476.0	137.0	102.2	80.6	388.8	83.8	235.0	132.3
UF 21333	Aucilla River, Florida	524.0	141.0	109.0	80.8	453.5	95.0	253.0	140.0
ANSP 12486	Big Bone Lick KY	470.0	140.0	107.5	71.9	382.4	~80	X	133.2
UCMP 21003	Rancho La Brea, California	457.0	128.0	102.0	75.3	373.5	82.3	232.0	127.5
UW20788	SeaTac Airport, Washington	509.9	139.8	112.3	87.3	400.0	78.8	235.9	129.9
DMNH G-25695	Darke County, Ohio	565.6	161.6	128.8	90.2	478.5	91.5	277.3	141.8
UF 24101	Warm Mineral Springs, Florida	498.0	120.2	93.9	64.3	413.0	95.0	195.0	118.9
UF 21333	Aucilla River, Florida	520.0	142.1	110.1	78.2	450.0	94.4	253.0	141.0
ChM PV7681	Cooper River, South Carolina	X	115.2	95.8	69.4	X	X	X	X

TABLE 7: Measurements of *Megalonyx* ulnae from Camelot L.F. to specimens from other localities (data from McDonald 1977); 1. Greatest length; 2. Length of olecranon process; 3. Transverse diameter of distal end; 4. Anteroposterior diameter of distal end; 5. Anteroposterior diameter across coronoid process; 6. Transverse diameter of trochlear notch; 7. Epiphyses fused? Summary statistics are included for Camelot samples.

Megalonyx from Camelot L.F. (SCSM)		1	2	3	4	5	6	7
2004.1.84		390.0	63.1	33.1	58.9	76.1	41.5	No
2003.75.365		X	X	32.5	43.7	X	X	No
2003.75.386		314.0	62.2	X	X	X	42.6	No
2003.75.387		334.0	X	33.9	43.0	X	X	Yes
2004.1.125		X	65.7	30.6	36.3	72.3	46.6	Yes
Camelot L.F. Summary Statistics	Mean	346.0	63.7	32.5	45.5	74.2	43.6	
	SD	39.4	1.85	1.4	9.5	2.71	2.67	
	n	3	3	4	4	2	3	
Megalonyx jeffersonii								
ISUM 23024	Amer. Falls Reservoir, Idaho	492.0	96.0	50.2	43.6	113.4	99.5	
ISUM 16462	Amer. Falls Reservoir, Idaho	45.0	96.5	44.3	44.2	138.4	114.7	
ANSP 12508	Cromer Cave, West Virginia	505.0	108.0	48.0	37.0	115.0	101.0	
UCMP 23192	Rancho La Brea, California	467.9	100.4	45.6	41.2	114.7	118.5	
UF 21335	Aucilla River, Florida	473.2	87.1	45.7	33.7	115.6	87.9	
UF 21336	Aucilla River, Florida	491.3	111.9	51.8	39.3	127.5	110.0	
UF 21337	Aucilla River, Florida	472.7	89.0	43.7	29.5	105.4	87.9	
UF 14974	Aucilla River, Florida	481.0	122.5	51.2	43.5	122.2	113.9	

TABLE 8. Measurements of *Megalonyx* radii from Camelot L.F. compared to various Rancholabrean specimens. Data from McDonald (1977), McDonald and Anderson (1983), McDonald and Ray (1990), McDonald et al. (2001): 1. Length of radius from head along posterior edge to end of lateral process; 2. Anteroposterior length of head across ulnar facet; 3. Mediolateral width of head parallel to bicipital tuberosity; 4. Anteroposterior width of midshaft; 5. Mediolateral width of distal end across medial and lateral processes; 6. Anteroposterior length of distal end from styloid process to ulnar edge; 7. Epiphyses fused? Summary statistics are included for Camelot samples.

Megalonyx from Camelot L.F. (SCSM)		1	2	3	4	5	6	7
2003.75.388		418.0	51.4	54.3	74.8	79.3	62.3	Yes
2004.1.85		327.0	X	X	66.8	X	X	No
2003.75.389		X	47.4	54.3	X	X	X	No
2003.75.390		X	46.4	54.6	X	X	X	No
2004.1.86		413	35.5+	46.19+	68.9	54.3	76.5	No
2003.75.440		X	X	X	X	58.7	77.8	No
2004.1		385.0	53.0	50.1	60.4	61.8	71.7	Yes
Camelot L.F. Summary Statistics	Mean	385.8	48.4	53.3	67.7	63.5	72.1	
	SD	41.8	2.6	2.1	5.9	11.0	7.0	
	n	4	4	4	4	4	4	

28

Megalonyx leptostomus							
USNM 23208	Hagerman, Idaho	X	44.1	46.7	65.0	X	X
UMMP V-57476	Hagerman, Idaho	X	X	X	67.3	45.2	68.0
UF 23527	Inglis IA, Florida	199.8	32.0	26.3	42.9	29.8	53.1
UF 23530	Inglis IA, Florida	X	40.5	38.1	X	X	X
UF 23528	Inglis IA, Florida	X	30.4	29.5	43.5	X	X
UF 23529	Inglis IA, Florida	X	31.4	25.6	39.0	X	X
Megalonyx wheatleyi							
F:AM 99195	McLeod, Florida	241.4	49.5	43.3	53.6	45.5	74.7
F:AM 105-1986	McLeod, Florida	359.8	49.0	47.5	77.1	63.7	72.1
F:AM 99196	McLeod, Florida	338.2	51.6	48.7	68.8	54.7	70.2
Megalonyx jeffersonii							
SUI 275	Turin L.F., IA	379.0	53.6	61.0	76.9	68.8	85.9
ISUM 16462	American Falls Reservoir, Idaho	370.8	58.9	60.8	95.2	77.5	92.2
ISUM 23034	American Falls Reservoir, Idaho	395.0	58.8	55.3	79.3	78.6	85.9
USNM 24591	Ladds Quarry, Georgia	372.3+	60	58.5	74.9	X	X
USNM 25175	Sandy Hook, New Jersey	X	53.0	51.0	X	X	88.0
BYU 802/13610	Point-of-the-Mountain, Utah	~435	66.8	66.7	X	86.2	122.5
DMNH G-25748	Darke County, Ohio	~450	76.4	64.0	X	~79	~107
ANSP 12508	Cromer Cave, West Virginia	396.0	51.0	59.0	79.0	79.0	83.0

not seem to be a discernable pattern in the measurements, and such variation is not known for these forelimb elements.

Carpals—Twelve carpal elements are present in the Camelot sample, representing all but two carpal bones (pisiform and falciform) of the manus. Carpal elements present include two lunars, two magna, two scaphoids, two trapezoids, three ulnares, and an unciform fused with a magnum, a pathological condition (Figure 11). Three of the carpals (SCSM 2007.75.442) are associated wrist bones of the right manus and include an ulnare, a trapezoid, and a lunar. Three other carpals (SCSM 2003.75.385) are associated with a metacarpal and five phalanges of the left manus (see sections on metacarpals and phalanges of the manus). Measurements of all elements are provided in Table 9.

Metacarpals—Twelve metacarpal elements are present, including one first metacarpal (MC I). As with most *Megalonyx* and some other ground sloths, the first metacarpal is co-ossified with the trapezium (McDonald 1977). There are five second metacarpals (MC II) in the Camelot sample that are distinguishable from other metacarpals by the laterally offset carina (McDonald 1977). One MC II is associated with three carpals and five phalanges (SCSM 2003.75.385). See the sections on carpals and phalanges of the manus for their respective measurements. There are two third metacarpals (MC III), each of which displays the typical stocky, symmetrical appearance of the element (McDonald 1977). Two forth metacarpals (MC IV), as discerned by their great length and round shaft and two fifth metacarpals (MC V) are also present. Measurements of metacarpals are presented in Table 10.

Phalanges of the manus—Numerous proximal, medial, and distal (ungual) phalanges were present in the Camelot *Megalonyx* sample, but not all of them could be assigned to a position on the manus or pes. The elements known from the manus are presented in Table 11. Associated manus elements (SCSM 2003.85.384) are pictured in Figure 12. Additionally, five phalanges are associated with one metacarpal and three carpal elements (SCSM 2003.75.385) of the left manus. See the sections on carpals and metacarpals for their measurements. Phalanges of unknown position are listed in Appendix 3.

Femur—There are two nearly complete femora, two partial femora, and five condylar epiphyses (femoral heads) in

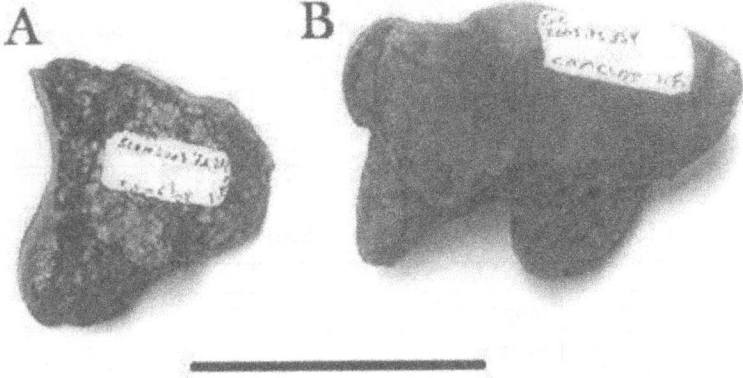

FIGURE 11. Carpal elements of *Megalonyx* from Camelot L.F. **A**, normal left magnum (SCSM 2003.75.371). **B**, pathological fusion of magnum with unciform (SCSM 2003.75.354). Scale bar = 5cm.

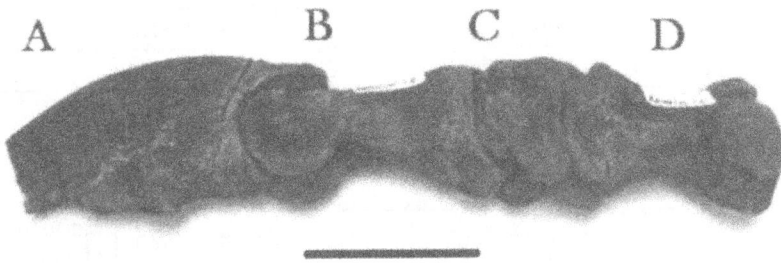

FIGURE 12. Associated elements of digit II of the right manus of *Megalonyx* from Camelot L.F. All elements are cataloged under SCSM 2003.75.384. From left to right: distal (ungual) phalanx, medial phalanx, proximal phalanx, metacarpal (MC II). Left lateral view. Scale bar = 5cm

TABLE 9. Measurements of *Megalonyx* carpals from Camelot L.F. (SCSM) compared to specimens from the late Rancholabrean Point-of-the-Mountain Utah (BYU): 1. Mediolateral width; 2. Dorsoventral depth; 3. Anteroposterior length. Summary statistics are included for Camelot samples with more than two specimens.

Catalog #	Element		1	2	3	Comments
SCSM 2003.75.385	lunar, left		47.8	34.26	28.6	
SCSM 2003.75.442.3	lunar, right		47.8	29.1	33.7	
BYU 802/13610	lunar, right		36.9	50.1	63.3	McDonald et al. (2001)
SCSM 2003.75.371	magnum, left		48.8	25.7	44.4	
SCSM 2003.75.369	magnum, right		42.4	22.4	40.2	
SCSM 2003.75.352	scaphoid		61.9	44.3	45.8	
SCSM 2003.75.366	scaphoid		60.0	41.8	50.3	
BYU 802/13610	scaphoid, right		77.5	71.5	52.5	McDonald et al. (2001)
SCSM 2003.75.385	trapezoid, left		33.7	21.8	34.0	
SCSM 2003.75.442.2	trapezoid, right		35.2	17.7	33.1	
SCSM 2003.75.385	ulnare, left		61.0	41.1	40.5	
SCSM 2003.75.351	ulnare, right		60.9	41.1	51.7	
SCSM 2003.75.442.1	ulnare, right		67.7	49.9	46.7	
Camelot L.F. Summary Statistics		Mean	63.2	44.0	46.3	
		SD	3.9	5.1	5.6	
		n	3	3	3	
BYU 802/13610	ulnare, right		80.8	52.4	62.1	McDonald et al. (2001)
SCSM 2003.75.354	unciform fused with magnum		65.0	30.9	38.5	PATH-OLOGY

32

TABLE 10. Measurements of *Megalonyx* metacarpals from Camelot L.F. (SCSM) compared to specimens from various Rancholabrean sites: Point of the Mountain, Utah (BYU); Darke County, Ohio (DMNH); Hunt County, Texas (USNM); Old Crow Basin, Yukon, Canada (CM); Aucilla River, Florida (UF). Data other than Camelot is from McDonald (2001): 1. Greatest length; 2. Mediolateral width of proximal end; 3. Dorso-ventral length of proximal end; 4. Mediolateral width of distal end; 5. Length of carina-distal end; 6. Epiphyses fused? Summary statistics are included for Camelot samples with more than two specimens.

Catalog #	Element (metacarpal)	1	2	3	4	5	6
SCSM 2003.75.449	MC I, right	42.8	21.7	32.4	18.1	24.6	Yes
BYU 802/13610	MC I, right	63.8	41.8	X	26.9	X	
SCSM 2003.75.384	MC II	75.1	37.8	39.1	30.3	46.2	No
SCSM 2003.75.356	MC II	69.3	43.1	43.8	33.3	51.0	Yes
SCSM 2004.1.57	MC II, right	77.2	47.9	47.0	34.8	X	No
SCSM 2003.75.385	MC II, left	75.1	38.1	33.8	27.9	40.4	No
SCSM 2003.75.362	MC II, right	89.2	42.1	49.7	34.0	54.3	Yes
Camelot L.F. Summary Statistics	Mean	77.2	41.8	42.7	32.0	48.0	
	SD	7.4	4.1	6.3	2.9	6.0	
	n	5	5	5	5	4	
BYU 802/13610	MC II, left	98.2	60.1	58.7	38.8	58.5	
DMNH G-25748	MC II, left	98.4	58.2	44.9	43.2	59.4	
USNM 10836	MC II, left	83.9	45.4	44.1	29.7	48.9	
SCSM 2003.75.361	MC III,	84.4	37.3	42.3	28.4	46.4	Yes
SCSM 2003.75.360	MC III,	81.1	34.1	42.0	29.1	46.1	Yes
BYU 802/13610	MC III, left	114.6	77.7	62.4	45.1	71.0	
CMN 48628	MC III, right	95.9	50.2	47.7	39.6	52.0	
UF 103604	MC III, right	105.8	64.1	61.4	44.8	61.3	
SCSM 2003.75.358	MC IV, left	106.7	35.1	43.3	33.4	53.0	Yes
SCSM 2003.75.359	MC IV, left	105.9	29.6	45.6	30.8	46.6	Yes
BYU 802/13610	MC IV, left	128.4	55.3	59.6	48.2	69.2	
SCSM 2003.75.376	MC V,	104.8	33.0	33.3	19.2	36.7	Yes
SCSM 2003.75.443	MC V,	92.7	30.5	31.6	33.0	X	No
BYU 802/13610	MC V, left	122.6	53.0	47.3	29.1	54.3	
DMNH G-25748	MC V, left	135.5	49.2	49.9	31.7	57.6	

TABLE 11. Measurements of *Megalonyx* phalanges of the manus from Camelot L.F.: 1. Maximum proximodistal depth; 2. Maximum mediolateral width; 3. Maximum anteroposterior length. Summary statistics are included for samples with more than two specimens.

Catalog # SCSM	Element		1	2	3
2003.75.337	phalanx-distal (digit I)		63.9	19.2	26.6
2003.75.340	phalanx-distal (digit I)		60.1	21.6	33.1
2004.1.81	phalanx-distal (digit I)		61	20.4	29.4
Summary Statistics for distal phalanx of digit I of the manus		Mean	61.7	19.8	28.9
		SD	2.0	1.2	3.2
		n	3	3	3
2003.75.370	phalanx-proximal (digit I)		45.9	18.4	25.4
2003.75.384	phalanx-proximal (digit II)		27.4	32.7	42.1
2003.75.384	phalanx-distal (digit II)		97.4+	23.0	45.4
2004.1.82	phalanx-distal (digit III)		150.6	39.7	68.7
2004.1.102	phalanx-distal (digit III)		142.7	36.0	61.8
2003.75.385	phalanx-medial (digit II manus)?		31.4	32.9	46.3
2003.75.384	phalanx-medial (digit II)		75.2	29.6	36.4
2003.75.385	phalanx-medial (digit III manus)?		66.1	37.1	41.3
2003.75.373	phalanx-medial (digit III)		75.1	40.6	47.4
2003.75.385	phalanx-proximal (digit III manus)?		32.7	38.3	50.7
2003.75.380	phalanx-proximal (digit III)		30.7	44.1	50.9
2003.75.385	phalanx-proximal (digit IV manus)?		30.6	32.2	44.5
2003.75.433	phalanx-proximal (digit V)		25.6	21.1	31.2
2003.75.433	phalanx-proximal (digit V)		24.6	19.4	29.9
2004.1.114	phalanx-proximal (digit V)		28.0	17.0	25.4
2004.1.133	phalanx-proximal (digit V)		23.8	21.8	29.1
Summary Statistics for proximal phalanx of digit V of the manus		Mean	25.5	19.8	28.9
		SD	1.8	2.1	2.5
		n	4	4	4

34

the Camelot material (Table 12). A partially reconstructed adult femur is shown in Figure 13. Note in Table 12 that total length of the Camelot specimens is larger than most other Irvingtonian *Megalonyx* (*M. wheatleyi*) and some Rancholabrean *Megalonyx* (*M. jeffersonii*). Because the femur is an important weight-bearing element of the skeleton, it is typically used estimations of body mass (Damuth and MacFadden 1990). Some investigators (McHenry 1992, Ruff et al. 1991, Grine et al. 1995) used the diameter of the condylar epiphysis (head) of the femur in body mass calculations. Roth (1990) used circumference of the femur, and Fariña (1998) used length of the femur to estimate body mass, See the section on body mass and Table 21 for complete discussion and the body mass estimates for Camelot *Megalonyx*.

Patella—There are two patellae known from the Camelot site (Figure 14). These are compared to *Megalonyx* patellae from other sites in Table 13. The late Irvingtonian Camelot specimens are ca. 20 mm smaller in most dimensions than late Rancholabrean *Megalonyx* from western localities.

Tibia—One complete adult tibia (Figure 15) and portions of two juvenile tibial elements are compared to other *Megalonyx* tibiae in Table 14. As with the femora, dimensions of Camelot *Megalonyx* tibiae, including juvenile specimens, are greater than most *M. wheatleyi* specimens and greater than or equal to some *M. jeffersonii* specimens. The *Megalonyx* tibiae from Camelot exhibit another trait that suggests that the population was in a state of transitional towards larger size. In another paper where evolutionary ratios and rates in *Megalonyx* are being calculated (Fields, unpublished data) I have noted a trend in which the proximal width of the tibia increases at an accelerated rate over time. This trend is especially noticeable from the late Irvingtonian (the time of the Camelot population) to the late Rancholabrean.

Fibula—One complete adult fibula (Figure 16) and portions of three juvenile fibulae are present. These are compared to other *Megalonyx* fibulae in Table 15. McDonald (1977) said that there was little change other than size in *Megalonyx* fibulae over the Plio-Pleistocene, although the Camelot sample is not significant to address this trend, the elements present and the other previously mentioned data (Fields, unpublished) seem to support McDonald's (1977) observation.

TABLE 12. Measurements of *Megalonyx* femora from Camelot L.F. compared to specimens from other localities (data from Gillette et al. 1999, McDonald and Anderson 1983, and Mills 1975): 1. Length along medial side from top of head to distal surface of medial condyle; 2. Greatest transverse diameter of proximal end; 3. Mediolateral diameter of head; 4. Anteroposterior diameter of head; 5. Mediolateral diameter of distal end across epicondyles; 6. Transverse diameter of shaft proximal to epicondyles and distal to third trochanter; 7. Epiphyses fused? Summary statistics are included for Camelot samples.

Megalonyx from Camelot L.F. (SCSM)			1	2	3	4	5	6	7
2004.1.92	femoral head		X	X	85.0	86.0	X	X	No
2003.75.395	femoral head		X	X	86.0	90.0	X	X	No
2003.75.396	femoral head		X	X	92.0	89.0	X	X	No
2003.75.397	femoral head		X	X	95.0	87.0	X	X	No
2003.75.398	femoral head		X	X	90.0	88.0	X	X	No
2004.1.97	femur, right		470.0	205.0	98.3	102.5	210.0	136.2	Yes
2004.1.96	femur, right		350+	180+	X	X	194+	107.0	No
2004.1.95	femur, left		350+	176+	X	X	185+	106.3	No
2004.1.126	femur, left		470.0	170+	97.0	102.1	200.0	133.6	Yes
Camelot L.F. Summary Statistics	Mean		470.0	205.0	91.9	92.1	205.0	120.8	
	SD		0.0	0.0	5.2	7.1	7.1	16.3	
	n		2	2	7	7	2	4	
Megalonyx leptostomus									
LACM 55828	femur	Ringhold, Washington	411.0	152.3	74.9	79.0	158.0	94.3	
UF 23546	femur-distal	Inglis 1A, Florida	X	X	X	X	139.8	76.8	Yes
UF 23547	femur-distal	Inglis 1A, Florida	X	X	X	X	144.1	X	Yes
UF 23795	femur-prox.	Inglis 1A, Florida	X	153.3	75.2	78.6	X	X	Yes

Megalonyx wheatleyi

F:AM 103-1986	femur	McLeod, Florida	440.0	183.0	80.4	87.4	188.0	124.3
F:AM 103-1920	femur	McLeod, Florida	417.7	X	85.3	X	201.5	X
UCMP 56088	femur	Irvington, California	500.0	~200	108.5	116.5	212.5	146.1

Megalonyx jeffersonii

USNM 64-R-15 6-7	femur	Valsequillo, Mexico	440.0	X	95.4	100.0	210.3	X	
SUI 48868	femur, right	Turin L.F., IA	563.0	245.0	116.0	116.0	255.0	X	
UF 21343	femur, right	Aucilla River, Florida	417+	214.0	100.4	102.7	X	146.2	Yes
IMNH 17060	femur	American Falls, Idaho	478.0	216.2	102.3	109.9	231.0	142.9	
IMNH 23034	femur	American Falls, Idaho	514.2	250.0	110.0	112.6	241.5	160.0	
BYU 13301	femur	Orem, Utah	505.0	225.2	115.1	108.6	224.7	~159	
UW 20788	femur	SeTac Washington	506.0	218.2	111.0	110.4	235.6	164.1	
USNM 23737	femur	Saltville, Virginia	525.6		106.7	112.0	238.2	X	
SMU 60247	femur	Moore Pit, Texas	413.3	X	83.2	82.0	184.0	X	

TABLE 13. Measurements of *Megalonyx* patellae from Camelot L.F. compared to specimens from late Rancholabrean localities (Gillette et al. 1999): 1. Dorsoventral diameter; 2. Transverse diameter; 3. Dorsoventral diameter of femoral articular facet; 4. Transverse diameter of femoral articular facet; 5. Thickness through femoral articular facet. Summary statistics are included for Camelot samples.

Catalog #	Element	Locality	1	2	3	4	5
SCSM 2004.1.105	patella, right	Camelot L.F., South Carolina	82.3	73.3	40.6	71.4	31.5
SCSM 2003.75.327	patella, left	Camelot L.F., South Carolina	86.8	75.3	45.5	70.0	31.3
BYUVP 13301	patella	Orem, Utah	105.1+	97.6+	60.6	89.2	51.7
UCMP 2277	patella	Rancho La Brea, California	106.0	99.0	52.0	97.6	42.2
LACM 6001-1	patella	Rancho La Brea, California	109.0	90.8	45.4	90.0	41.0

TABLE 14. Measurements of *Megalonyx* tibiae from Camelot L.F. compared to specimens from other localities (Gillette et al. 1999): 1. Length from top of intercondylar spine to middle of astragalar surface; 2. Least width of shaft; 3. Mediolateral diameter of distal end; 4. Mediolateral diameter of proximal end from edge of medial condyle to tibial crest; 5. Anteroposterior diameter of proximal end from posterior edge of lateral condyle to anterior edge of medial condyle; 6. Epiphyses fused? Summary statistics are included for Camelot samples.

SCSM No.	Element	Locality	1	2	3	4	5	6
2003.75.403	tibia, left	Camelot L.F., South Carolina	297.0	62.9	141.1	174.8	118.7	Yes
2004.1.80	tibia, right		232+	53.4	122.0	X	X	No
2003.75.450	distal epiphysis		X	X	128.3	X	X	No
Megalonyx leptostomus								
PPHM 1716	tibia	Cita Canyon, Texas	237.6	47.2	104.4	134.0	X	
WTSU (no #)	tibia	Cita Canyon, Texas	X	55.2	X	140.0	X	
UF 21348	tibia	Inglis 1A, Citrus Co., Florida	197.2	44.9	83.6	X	X	Yes
Megalonyx wheatleyi								
ANSP 15554	tibia	Port Kennedy Cave, Pennsylvania	287.1	61.0	119.3	~129	116.0	
ANSP 15555	tibia	Port Kennedy Cave, Pennsylvania	268.7	62.0	123.3	166.9	X	
ANSP 186	tibia	Port Kennedy Cave, Pennsylvania	257.9	64.5	116.6	X	X	
F:AM 103-1986	tibia	McLeod, Florida	247.2	62.8	118.2	154.1	162.6	
Megalonyx jeffersonii								
IMNH 149	tibia	Amer. Falls Reservoir, Idaho	345.1	80.1	155.9	211.5	X	
TMM 30967-1231	tibia	Ingleside, Texas	272.8	62.7	128.5	~173	157.0	
MR 002	tibia	Doedon Local Fauna, Montana	X	78.8	156.8	192.4	189.6	
BYUVP 13301	tibia	Orem, Utah	~303	81.3	149.4	~172	177.3	
IMNH 1781	tibia	Acequia Gravel Pit, Idaho	333.1	88.0	163.1	198.0	194.1	
UW 20788	tibia	SeTac Airport, Washington	282.6	71.5	142.6	184.6	201.0	
UF 23569	tibia	Warm Mineral Springs, Florida	276.2	66.6	123.5	175.9	182.0	

TABLE 15. Measurements of *Megalonyx* fibulae from Camelot L.F. compared to specimens from other localities (Hirschfeld and Webb 1968 and McDonald 1977): 1. Greatest length; 2. Anteroposterior width of proximal end across articular facet; 3. Anteroposterior width of distal end across articular facet; 4. Mediolateral width of proximal end across anterior surface; 5. Mediolateral width of distal end across anterior surface; 6. Epiphyses fused? Summary statistics are included for measurement #1 (greatest length) of the Camelot sample.

Megalonyx from Camelot L.F. (SCSM)		1	2	3	4	5	6
2004.1.89		280.0	65.0	68.8	63.4	63.9	Yes
2004.1.118		X	X	56.8	X	65.4	No
2004.1.87		210.0	X	X	X	X	No
2004.1.88		207.0	X	X	X	X	No
Camelot L.F. Summary Statistics	Mean	232.3					
	SD	41.3					
	n	3					
Megalonyx leptostomus							
USNM 23209	Hagerman, Idaho	280.0	X	63.3	X	63.8	
UF 10354	Mabel, Florida	221.5	41.6	44.0	39.2	50.8	
UF 23467	Inglis IA, Florida	193.3	36.1	42.7	31.8	41.5	
Megalonyx wheatleyi							
F:AM 103-1986	McLeod, Florida	281.5	61.0	57.4	55.5	66.2	
Megalonyx jeffersonii							
UW 20788	Seattle Tacoma Airport, Washington	302.5	65.1	81.4	63.4	85.8	
UF 14974	Aucilla River, Florida	328.0	69.0	79.2	73.9	87.5	
UF 14888	Aucilla River, Florida	313.6	57.4	70.8	64.4	75.1	
DMNH G-25930	Darke County, Ohio	384.0	82.9	92.8	82.5	98.5	

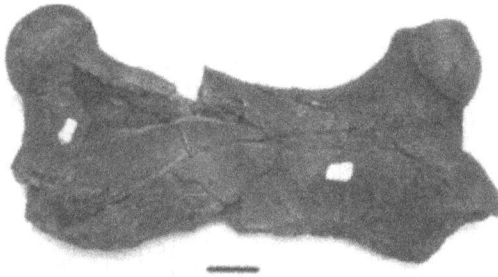

FIGURE 13. Lateral view of partially reconstructed left femur of *Megalonyx* (SCSM 2004.1.26) from Camelot L.F. Scale bar = 5cm.

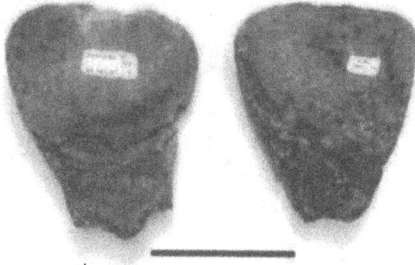

FIGURE 14. Patellae of *Megalonyx* from Camelot L.F.: Left patella (SCSM 2003.75.327), anterior view, and right patella (SCSM 2004.1.105), anterior view. Scale bar = 5cm.

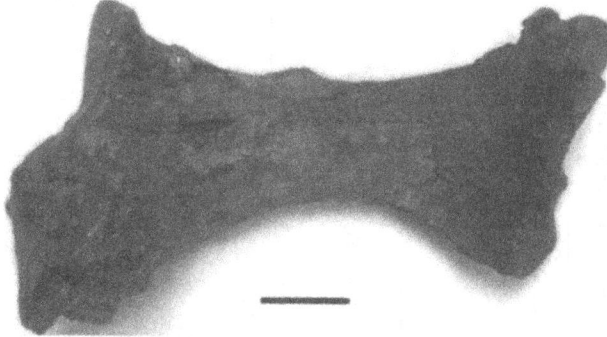

FIGURE 15. Left tibia of *Megalonyx* (SCSM 2003.75.403), left lateral view from Camelot L.F. Scale bar = 5cm.

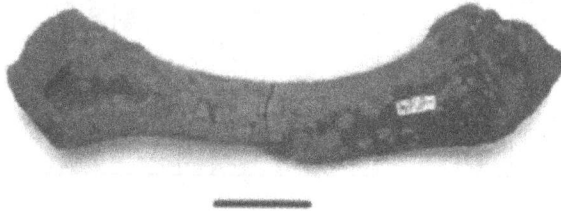

FIGURE 16. Right fibula of *Megalonyx* (SCSM 2004.1.89) from Camelot L.F. Scale bar = 5cm.

Astragalus—There are two left astragali present from the Camelot site. Specimens are pictured in Figure 17 and listed in Table 16. Both of these specimens exhibit typical features of the *Megalonyx* astragalus, including the unmodified medial trochlea and concave articular surface (McDonald et al. 2000). As with other Camelot elements, several astragalar dimensions are greater than those of *M. wheatleyi* and greater than or equal to those of *M. jeffersonii*.

Calcaneum—There are seven calcanea (four adult and three juvenile) from the Camelot site. (Figure 18, Table 17). Again, measurements of Camelot *Megalonyx* suggest size between a large *M. wheatleyi* to a small *M. jeffersonii*. See the section on taxonomy for further discussion of the implications of calcanear dimensions.

Tarsals—Seven tarsal elements were recovered from Camelot. Measurements for two naviculars are included in Table 18. No other tarsal elements from other localities were available for comparative measurements. The other tarsal elements from the Camelot site include a cyamelle (SCSM 2003.75.353), two ectocuneiforms (SCSM 2003.75.382, 2004.1.109), a cuboid (SCSM 2003.75.452.3) and a mesocuneiform (SCSM 2003.75.452.4).

Metatarsals—Nine metatarsal elements representing all five digits of the pes are present in the Camelot sample: three first metatarsals (MT I); one second metatarsal (MT II); one third metatarsal (MT III); one fourth metatarsal (MT IV); three fifth metatarsals (MT V). See Table 19 for measurements.

Phalanges—At least 30 proximal, medial, and distal (ungual) phalanges were present, but not all of them could be assigned to a position on the manus or pes. The elements known from the pes are presented in Table 20. Phalanges of unknown position are presented in Appendix 3.

Within the genus *Megalonyx* pedal phalanges are used as a qualitative character to distinguish *M. wheatleyi* from *M. jeffersonii*. According to McDonald (1977) proximal and medial phalanges of the third pedal digit that are not ossified indicate *M. wheatleyi*; if these phalanges are co-ossified, then the specimen is referred to *M. jeffersonii*. The Camelot sample includes associated proximal (SCSM 2004.1.131) and medial (SCSM 2004.1.130) phalanges of the third digit of the pes that are separate and distinct (see Figure 19). According to McDonald (1977) this character identifies the Camelot

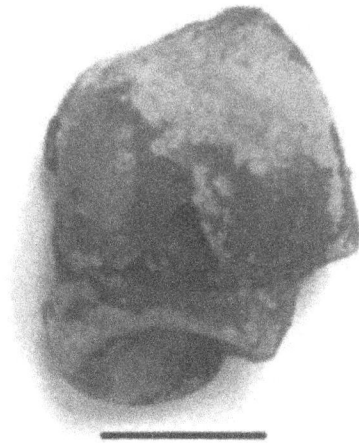

FIGURE 17. Dorsal view of left astragalus (SCSM 2003.75.383) of *Megalonyx* (dorsal view) from Camelot L.F. Scale bar = 5cm.

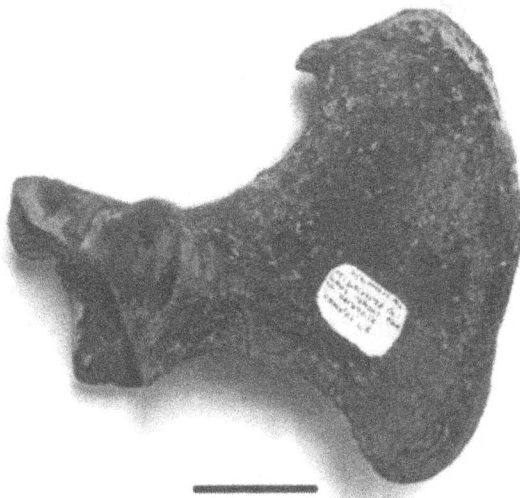

FIGURE 18. Left calcaneum (SCSM 2003.75.392) of *Megalonyx* (plantar view) from Camelot L.F. Scale bar = 5cm.

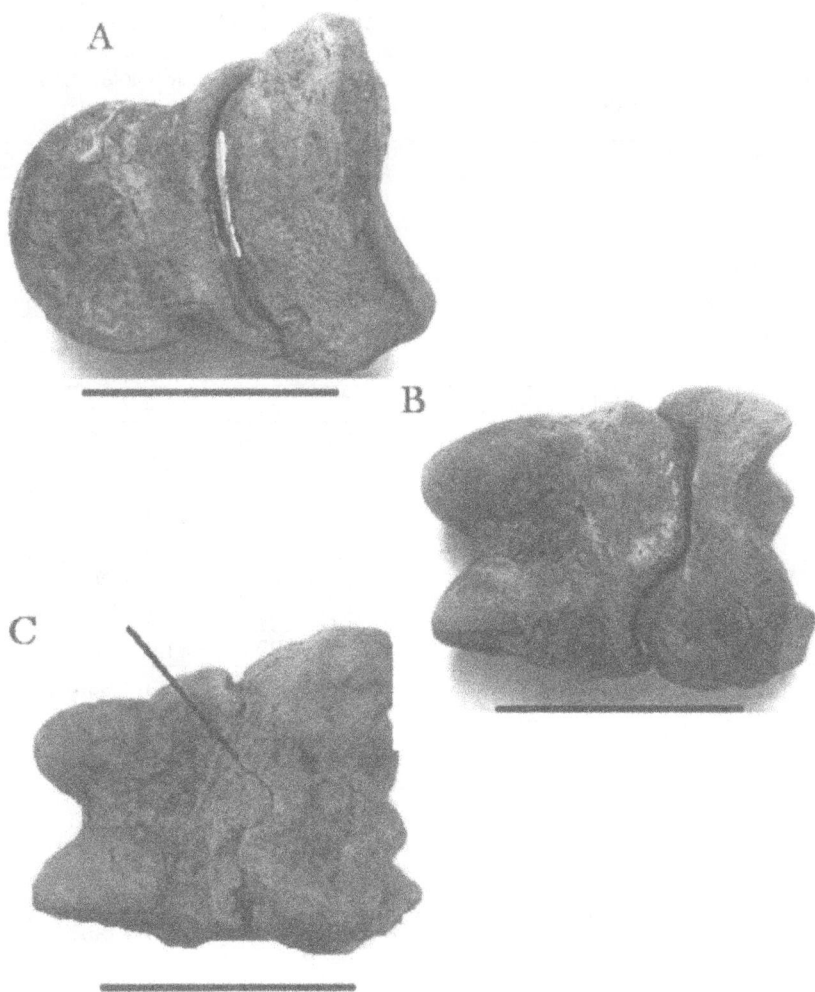

FIGURE 19. Unfused phalanges of digit III of the pes in *Megalonyx* from the Camelot Local Fauna. **A**, Medial (SCSM 2004.1.130) and proximal (SCSM 2004.1.131) phalanges, lateral view. **B**, dorsal view. **C**, ventral view showing suture line. **D**, articular surfaces of each element showing small protuberances of bone. Scale bar in each image = 5cm.

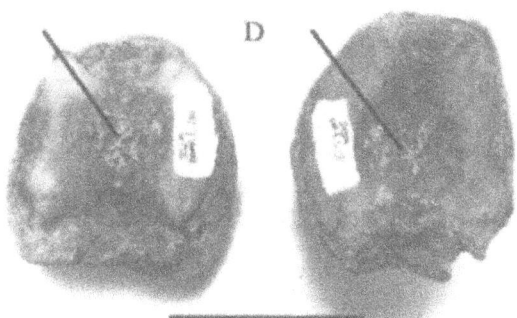

TABLE 16. Measurements of *Megalonyx* astragali from Camelot L.F. compared to specimens from other localities (data from McDonald 1977, McDonald et al. 2000, and McDonald et al. 2001): 1. Anteroposterior length from middle of trochlea (posterior edge) to anterior surface of navicular process; 2. Mediolateral width of posterior edge of trochlea; 3. Mediolateral width of anterior edge of trochlea; 4. Mediolateral width of navicular process; 5. Dorsoventral height of navicular process.

Megalonyx from Camelot L.F. (SCSM)		1	2	3	4	5
2003.75.383		106.1	60.0	81.1	54.1	46.5
2003.75.452.1		110.9	63.5	81.7	58.3	47.0
Megalonyx leptostomus						
USNM 23209	Hagerman, Idaho	83.2	X	67.8	59.6	39.3
UF 21407	Haile 16A, Florida	95.7	54.0	66.1	43.7	32.8
UF 21409	Inglis 1A, Florida	70.3	41.8	53.5	33.1	24.4
UF 21410	Inglis 1A, Florida	62.7	36.4	45.4	31.0	22.7
UF 21410	Inglis IA, Florida	59.0	34.6	47.2	40.4	29.1
UF ML-9	Haile 7G, Florida	71.9	46.5	56.6	43.5	31.0
UF ML-8	Haile 7G, Florida	69.4	45.6	55.4	39.5	27.9
UF ML-14	Haile 7G, Florida	76.5	54.6	59.3	50.8	33.8
UF ML-10	Haile 7G, Florida	70.4	47.7	62.0	36.6	25.4
UF ML-11	Haile 7G, Florida	72.1	49.2	61.4	34.0	25.6
UF 21409	Inglis 1A, Florida	68.7	39.0	51.9	36.4	24.0
UF 10458-A	Santa Fe River I, Florida	54.7	32.5	43.0	38.1	30.6
Megalonyx wheatleyi						
LACM 1377/3634	Vallecito, California	86.4	57.7	70.9	63.4	49.8
ANSP 15555	Port Kennedy, Pennsylvania	80.5	51.6	66.8	60.6	42.5
ANSP 15558	Port Kennedy, Pennsylvania	97.3	64.5	85.5	~70.6	52.4
ANSP 15557	Port Kennedy, Pennsylvania	91.2	66.9	74.0	66.8	50.9
ANSP 15560	Port Kennedy, Pennsylvania	94.8	64.7	78.3	72.4	48.2
F:AM 102-1938	McLeod, Florida	82.2	57.1	67.7	58.6	41.1
UF 21407	Haile 16, Florida	75.1	51.7	65.4	50.5	39.4
Megalonyx jeffersonii						
USNM 424556	Edisto Island, South Carolina	109.7	71.8	76.8	58.0	44.9
ISUM 17122	American Falls Reservoir, Idaho	101.6	72.0	86.3	81.1	60.5
ISUM 28288	American Falls Reservoir, Idaho	100.5	82.3	93.1	X	61.3
ISUM 17288	American Falls Reservoir, Idaho	101.6	72.1	90.7	74.6	56.8
UF 14888	Aucilla River, Florida	109.0	72.6	92.8	67.1	54.0
CMN 25148	Old Crow Basin, Canada	102.2	71.9	83.2	69.7	48.4
NMC 25148	Old Crow River, Yukon	101.4	71.7	84.0	69.4	47.2
BYU 802/13610	Point-of-the-Mountain, Utah	92.9	120.5	~80	70.7	102.8
UW 20788	Seattle Tacoma Airport, Washington	104.9	73.2	90.0	75.3	61.5
UF 47961	Aucilla River, Florida	112.3	85.1	86.2	55.9	39.6

TABLE 17. Measurements of *Megalonyx* calcanea from Camelot L.F. compared to specimens from other localities (data from McDonald 1977): 1. Length from the posterior edge of the tuber calcis to the anterior edge of the sustentaculum; 2. Width of the tuber calcis; 3. Width of the neck; 4. Transverse length of the posterior astragalar facet; 5. Height of the proximal end from the dorsal edge of the posterior astragalar to the ventral edge of the cuboid process; 6. Epiphyses fused? Summary statistics are included for Camelot samples.

Megalonyx from Camelot L.F. (SCSM)		1	2	3	4	5	6
2003.75.393		209.0	185.0	60.3	52.2	92.8	Yes
2003.75.392		210.0	194.0	61.1	55.1	94.1	Yes
2004.1.136		209.0	200.0	68.4	63.3	84.1	Yes
2003.75.407		200.0	199.0	64.1	45.5	89.7	Yes
2003.75.394		164.6	155.9	58.2	42.6	82.6	No
2004.1.90		159.6	145.1	48.1	49.4	79.4	No
2004.1.91		158.3	144.9	47.9	48.9	79.8	No
Camelot L.F. Summary Statistics	Mean	187.2	174.8	58.3	51.0	86.1	
	SD	25.0	25.3	7.7	6.8	6.1	
	n	7	7	7	7	7	
Megalonyx leptostomus							
UF 21373	Haile 16A, Florida	X	X	44.8	27.8	71.1	
UF 21377	Inglis 1A, Citrus Co., Florida	145.3	140.9	40.2	39.5	59.5	Yes
UF 21376	Inglis 1A, Citrus Co., Florida	133.1	114.3	40.1	33.5	55.1	Yes
UF 21375	Inglis 1A, Citrus Co., Florida	124.4	111.4	37.0	36.9	54.7	Yes
USNM 23209	Hagerman, Idaho	167.4	164.5	47.8	49.4	83.6	
PPHM 2599	Cita Canyon, Texas	153.1	~139.4	44.5	47.0	73.2	
UF 21414	Santa Fe River I, Florida	X	~120	52.9	60.0	78.5	
UF 10349	Mable, Florida	118.4	133.9	42.2	40.8	65.3	
Megalonyx wheatleyi							
ANSP 269	Port Kennedy Cave, Pennsylvania	X	113.3	44.9	46.0	X	No
F:AM 99200	McLeod Limerock Mine, Florida	X	X	65.0	42.5	75.7	
F:AM 103-1986	McLeod Limerock Mine, Florida	187.0	193.1	54.9	50.5	77.6	
Megalonyx jeffersonii							
UF 103493	Aucilla River 3J (Page/Ladson), Florida	250.0	217.0	65.0	70.1	110.4	Yes
ChM PV5585	Edisto Beach, South Carolina	X	X	69.6	60.5	91.0	
ISUM 1535	America Falls Reservoir, Idaho	230.0	243.1	75.7	76.2	111.8	
UCMP 20095	Rancho La Brea, California	207.8	201.3	71.0	75.0	95.1	
DMNH G-25914	Darke County, Ohio	266.6	273.5	78.1	89.1	113.8	
USNM 23734	Lane Cave, Virginia	195.1	200.6	81.8	45.0	101.0	Yes

TABLE 18. Measurements of *Megalonyx* naviculars from Camelot L.F. compared to specimens from other localities (Hirschfeld and Webb 1968, McDonald et al. 2001):1. Dorsoventral length; 2. Mediolateral width; 3. Anteroposterior thickness; 4. Length of ectocuneiform facet.

Megalonyx from Camelot L.F. (SCSM)	Element	Locality	1	2	3	4
2004.1.56	navicular		62.6	57.8	31.5	46.7
2003.75.452.2	navicular-left		69.5	60.2	29.6	50.3
Megalonyx leptostomus						
USNM 23209	navicular	Hagerman, Idaho	60.1	53.0	27.6	X
Megalonyx jeffersonii						
DMNH G-25748	navicular	Darke County, Ohio	90.2	81.3	45.2	85.5
USNM 214932	navicular	Melbourne, Florida	88.5	67.9	40.3	70.3

TABLE 19. Measurements of *Megalonyx* metatarsals from Camelot L.F. compared to specimens (where available) from other localities. *M. leptostomus* data from Hirschfeld and Webb (1968). Some *M. jeffersonii* data from Simpson (1928): 1. Greatest length; 2. Mediolateral width of proximal end; 3. Dorsoventral length of proximal end; 4. Mediolateral width of distal end; 5. Length of carina: distal end; 6. Epiphyses fused? Summary statistics are included for Camelot samples with more than two specimens.

Catalog #	Element	Locality	1	2	3	4	5	6
SCSM 2004.1.119	MT I	Camelot L.F., South Carolina	46.9	20.4	20.9	49.6	X	Yes
SCSM 2003.75.368	MT I, left	Camelot L.F., South Carolina	47.6	26.3	23.9	49.4	X	Yes
SCSM 2003.75.367	MT I, right	Camelot L.F., South Carolina	45.2	21.7	28.2	49.7	50.2	Yes
Camelot L.F. Summary Statistics		Mean	46.6	22.8	24.3	49.6	50.2	
		SD	1.3	3.1	3.7	0.2	X	
		n	3	3	3	3	1	
SCSM 2004.1.113	MT II	Camelot L.F., South Carolina	70.3	42.6	51.0	35.9	50.5	Yes

Specimen	Element	Locality						
M. leptostomus WT (no #)	MT II	Cita Canyon, Texas	54.0	X	38.0	37.0	X	
M. leptostomus USNM 23209	MT II	Hagerman, Idaho	35.0	X	39.0	X	29.7	
M. jeffersonii AMNH 23424	MT II, left	Citrus Co., Florida	67.0	X	46.0	X	49.0	
SCSM 2004.1.100	MT III	Camelot L.F., South Carolina	78.6	48.8	59.0	53.2	68.4	Yes
M. leptostomus WT (no #)	MT III	Cita Canyon, Texas	51.5	38.5	X	X	55.0	
M. leptostomus USNM 23209	MT III	Hagerman, Idaho	X	21.0	X	X	66.0	
M. jeffersonii UF 21463	MT III, right	Aucilla River, Florida	100.4	49.2	68.3	41.8	63.3	Yes
SCSM 2003.75.363	MT IV, right	Camelot L.F., South Carolina	100.5	43.0	59.5	36.2	57.1	Yes
M. leptostomus USNM 23209	MT IV	Hagerman, Idaho	X	41.0	X	X	50.1	
M. jeffersonii AMNH 23423	MT IV	Citrus Co., Florida	91.0	X	X	59.0	52.0	
M. jeffersonii UCMP 6002-1	MT IV, right	Rancho La Brea, California	93.4	49.4	65.0	41.6	63.3	
SCSM 2004.1.108	MT V	Camelot L.F., South Carolina	109+	35.5	46.3	35.5	52.8	No
SCSM 2004.1.59	MT V, right	Camelot L.F., South Carolina	117.4	40.2	49.8	38.2	65.5	Yes
SCSM 2004.1.132	MT V, right	Camelot L.F., South Carolina	116.9	36.0	46.5	42.8	53.1	No
Camelot L.F. Summary Statistics		Mean	117.2	37.3	47.5	38.8	57.1	
		SD	0.4	2.6	1.9	3.7	7.3	
		n	2	3	3	3	3	
M. leptostomus UF 21868	MT V, left	Inglis 1A, Florida	49.2	24.5	24.2	90.1	X	
M. leptostomus USNM 23209	MT V	Hagerman, Idaho	X	21.1	X	X	24.1	

TABLE 20. Measurements of *Megalonyx* phalanges of the pes from Camelot L.F.: 1. Maximum proximodistal depth; 2. Maximum mediolateral width; 3. Maximum anteroposterior length; 4. Epiphyses fused?

Catalog # SCSM	Element	1	2	3	4
2003.75.451	phalanx-distal (digit I) pes	47.9	25.6	36.1	Yes
2003.75.336	phalanx-distal (digit I) pes	39.5	21.3	30.2	Yes
2003.75.341	phalanx-distal (digit III) pes	~166	36.8	74.2	Yes
2003.75.439	phalanx-distal (digit III) pes	~167	46.5	74.0	Yes
2003.75.348	phalanx-medial (digit III) pes	62.0	48.3	54.4	Yes
2003.75.374	phalanx-medial (digit IV pes?)	72.4	39.9	37.1	Yes
2003.75.364	phalanx-medial (digit IV) pes	65.3	37.3	40.3	Yes
2003.75.435	phalanx-medial (pes?)	70.3	30.8	47.8	Yes
2003.75.436	phalanx-medial (pes?)	69.0	32.8	47.8	Yes
2004.1.54	phalanx-medial (digit III) pes	65.6	53.3	54.1	Yes
2004.1.131	phalanx-medial (digit III) pes	54.7	48.5	52.8	Yes
2004.1.130	phalanx-proximal (digit III) pes	30.1	53.0	60.9	Yes
2003.75.447	phalanx-proximal (digit V) pes	23.7	18.1	27.0	Yes

Megalonyx as *M. wheatleyi*. However, early signs of ossification are visible on the articular surfaces of the phalanges (see Figure 19).

Taphonomy

Remains of the animals comprising the Camelot Local Fauna may have been deposited as the result of two large flooding events, as might be associated with major hurricanes (J. Knight, South Carolina State Museum, pers. comm.). This working hypothesis may describe a scenario similar to what occurred in August 2005, when Hurricane Katrina swept through the Gulf Coast states, and was followed weeks later by Hurricane Rita.

According to this hypothesis, a major flood event in the region caused the deaths of most animals in the vicinity. The first flood, then, might have killed the animals and subsequent decomposition could have created a "float and bloat" scenario that would have deposited the carcasses in the lower areas where water carried them as suggested by the fact that the fossil remains were recovered from a channel fill. The bodies evidently were exposed to the atmosphere and to decomposers long enough for the stronger ligaments and tendons to deteriorate and allow bones to become separated.

The number of bones of juvenile *Megalonyx* and those of several other species in the Camelot sample suggests that young animals were unable to deal with high water. Of the 253 cataloged elements of *Megalonyx* from Camelot, there were 152 postcranials. Of those 152 elements, 46 (30.3%) had unfused epiphyses (i.e. represented juveniles). This number is in addition to the 30 vertebral centra with one or both epiphyses unfused. However, as stated earlier, fusion of epiphyseal plates to vertebral centra is not an indicator of adulthood in *Megalonyx* (McDonald 1977:93). Nevertheless, almost one-third of the Camelot sample is composed of immature individuals. In addition to immature *Megalonyx*, a number of old animals of other species, based on tooth wear, also were heavily selected against.

The second flood event would have had enough flow to orient the detached bones hydraulically. The orientation of the bones suggests that the channel from which they were excavated flowed from west to east, as does the river which currently drains the large swamp just north of the dig site. The condition of the bones, including nearly pristine unbroken limb bones and relatively intact skulls of several taxa, suggest that the fossils were not transported a great distance.

Specimens of the largest members of the megafauna, which should be present in the collection, are noticeably absent from the Camelot sample. No remains of proboscideans or giant ground sloths of the genus *Eremotherium* have been collected from the site. It seems likely that these organisms of extremely large body size may, if they were present, have been able to cope with the high water. The scarcity of fossils of large animals of reproductive age—usually healthy, experienced individuals—would appear to support that hypothesis.

Body Mass

Body mass has profound implications for many aspects of an animals' biology, and numerous ways to estimate body mass from skeletal features have proven useful in applications to fossil specimens (Damuth and MacFadden 1990). Body mass estimates for various extinct ground sloths were reported by Fariña et al. (1998), Christiansen and Fariña (2003), and McDonald (2005). These authors emphasized the importance of scaling and allometry in selecting appropriate skeletal elements and formulas for estimating body mass of extinct taxa.

In an effort to estimate body mass of Camelot *Megalonyx* I compared four sets of bones, measurements, and accompanying body mass formulas (Table 21). As most ground sloths, including *Megalonyx*, could adopt a bipedal stance and perhaps amble along on the hind legs, it might seem appropriate to use a body mass proxy for other bipedal organisms, such as humans. However, because bones of humans are scaled differently than ground sloths, such estimates are suspect. Indeed, use of the measurements of the femoral head with anthropological formulas (McHenry 1992, Ruff et al. 1991, Grine et al. 1995) resulted in estimates of body mass for *Megalonyx* that appeared quite low (ranging from 144.9 kg to 196.0 kg). Similar logic might suggest that since *Megalonyx* was part of the megafauna it would seem appropriate to use a body mass formula for another large mammal, such as an elephant. However, because proboscidean allometry differs significantly from that of pilosans, a formula using circumfer-

TABLE 21. Body mass estimates for Camelot L.F. *Megalonyx*. Abbreviations are: FHD = femur head diameter; FL = femoral length; FC = femur circumference; HC = humerus circumference. Bold values are closest to other published estimates of body mass for *Megalonyx* from the Late Pleistocene Irvingtonian.

Cat # (SCSM)	Element	FHD (mm)	FL (cm)	FC (mm)	HC (mm)	Epiphyses fused?	Body mass estimates (kg)				
							Ruff et al. 1991	McHenry 1992	Grine et al. 1995	Fariña et al. 1998	Roth 1990
2004.1.92	condylar epiphysis (head) of femur	86.0	X	X	X	No	144.9	152.7	158.5	X	X
2003.75.395	condylar epiphysis (head) of femur	90.0	X	X	X	No	152.6	161.6	167.6	X	X
2003.75.396	condylar epiphysis (head) of femur	89.0	X	X	X	No	150.7	159.4	165.4	X	X
2003.75.397	condylar epiphysis (head) of femur	87.0	X	X	X	No	146.8	154.9	160.8	X	X
2003.75.398	condylar epiphysis (head) of femur	88.0	X	X	X	No	148.8	157.1	163.1	X	X
2004.1.97	femur, right	102.5	47.0	317.5	X	Yes	177.0	189.6	196.0	**824.3**	4477.8
2004.1.126	femur, left	102.1	47.0	330.2	X	Yes	176.2	188.8	195.2	**824.3**	5002.9
2004.1.96	femur, right	X	X	266.7	X	No	X	X	X	X	2735.3
2004.1.95	femur, left	X	X	264.2	X	No	X	X	X	X	2663.4
2003.75.405	femur-partial	X	X	241.3	X	No	X	X	X	X	2061.2
2003.75.404	humerus-right	X	X	X	184.0	No	X	X	X	X	**775.5**
2004.1.112	humerus	X	X	X	171.0	No	X	X	X	X	**638.7**

51

ence of the femur (Roth 1990) resulted in apparent overesti-
mates of body mass (ranging from 2061.2 kg to 5002.9 kg) in
Megalonyx. However, while the femora of most ground sloths
are proportionally more compact than those of elephants,
the relatively long ground sloth forelimbs approach the same
scale as proboscideans. Using a formula that utilizes humerus
circumference (Roth 1990) I obtained estimates of 638.7 kg
and 775.5 kg, respectively, for two immature *Megalonyx* at
Camelot. Such results appear to be better estimates of body
mass as McDonald (2005) reported a similar mean estimate
(595.93 kg) for Irvingtonian *Megalonyx*.

Fariña et al. (1998) used a formula appropriate to xenarthran anatomy in body mass estimates of several pilosan
and cingulate taxa. McDonald (2005) used the same formula
to estimate body mass for 19 taxa of xenarthrans. I applied
this formula utilizing femur length, which resulted in a body
mass estimate of 824.3 kg. This value is between the 595.93
kg estimate for *M. wheatleyi* and the 1090.35 kg estimate for
M. jeffersonii (McDonald 2005). Although this estimate was
only attainable for two specimens from the Camelot site,
it appears to reinforce the notion that the Camelot sample
represents a transitional population of late Irvingtonian
Megalonyx.

Taxonomy

To diagnose *M. wheatleyi*, McDonald (1977:267–268) said of
the teeth, "upper caniniform with medial lingual bulge but
anterior and posterior longitudinal grooves weak or absent,"
and of the pes, "proximal and medial phalanx of third pedal
digit separate and distinct." In contrast, *M. jeffersonii* can be
identified by the "upper caniniform with prominent lingual
bulge medially located and with well developed anterior and
posterior longitudinal grooves." Additionally, the "proximal
and medial phalanges of third digit of pes [are] co-ossified
and the "tuber calcis is generally wider than length of calcaneum." (McDonald 1977: 257).

The Camelot *Megalonyx* appear to be a transitional form,
displaying characters of both *M. wheatleyi* and *M. jeffersonii*
as described by McDonald (1977). *Megalonyx* from Camelot
have upper caniniforms that possess a lingual bulge with
varying degrees of prominence, and longitudinal grooves are

not consistently well developed (see Figure 4). In addition, phalanges of the third pedal digit are not co-ossified (see Figure 19). Furthermore, none of the Camelot specimens possess a calcaneum in which the tuber calcis is wider than the total length (see Table 17: measurements 1 and 2). In fact, one calcaneum (SCSM 2003.75.407) is just one millimeter short of being equal in width and length. Moreover, two of the six calcanea identified as *M. jeffersonii* listed in Table 17 (UF 103493 and UCMP 20095) do not have a tuber calcis that is wider than the total length of the specimen.

Numerous measurements of postcranial elements from Camelot *Megalonyx* were larger than those of *M. wheatleyi* and in some cases as large as or larger than those of *M. jeffersonii* (Table 22). Of these elements 67.9% of the Camelot sample measurements are greater than those of *M. wheatleyi* surveyed from other sites. In addition, over 15% of the Camelot sample measurements are greater than or equal to those surveyed for *M. jeffersonii*.

Recognition of species in the fossil record can be difficult for a number of reasons, a notion that I discuss at length in another paper, as have several other authors (Benton and Pearson 2001, Bruner 2004, DeQuieroz 2005, Forey et al. 2004, Gingerich 1985). The difficulty of recognizing fossil species will not be discussed here, but it seems clear from the preceding qualitative and quantitative data that the Camelot sample of *Megalonyx* is a prime example of this problem. Having studied the material from Camelot, H.G. McDonald (pers. comm.) referred this sample to *M. wheatleyi* based on the qualitative characters. Some of these characters, however, are relative in nature and their recognition somewhat subjective i.e. prominence of lingual bulge and presence of grooves on upper caniniforms (see Figure 4). Moreover, the quantitative data, including significant overlap in caniniform measurements (Figure 3) and postcranial material (Table 22) make assignment of the Camelot *Megalonyx* to species even more problematic.

Rose and Bown (1986) noted a similar situation in Eocene onomyiid primates where the fossil record was sufficient to show transitions but no clear species boundaries. Thus, the assignment of species becomes arbitrary in the context of a well documented fossil record where gradual evolution is evident. However, as George Gaylord Simpson (1961:117)

TABLE 22. Overview of Camelot L.F. *Megalonyx* measurements compared to those of *M. wheatleyi* and *M. jeffersonii* presented in this paper. For each element listed, the number of Camelot measurements greater than those of *M. wheatleyi* and the corresponding percentage is given. Then, the number of Camelot measurements greater than or equal to those of *M. jeffersonii* and the corresponding percentage is given.

[1] Mean Camelot sample compared to individuals of *M. wheatleyi* and *M. jeffersonii*.
[2] Individual Camelot specimens (all juveniles) compared to individuals of *M. wheatleyi* and *M. jeffersonii*.
[3] Mean Camelot sample (including juveniles) compared to individuals of *M. wheatleyi* and *M. jeffersonii*.
[4] One adult from Camelot sample compared to individuals of *M. wheatleyi* and *M. jeffersonii*.

Element	Camelot measurements > *M. wheatleyi*		Camelot measurements ≥ *M. jeffersonii*	
Mandible[1]	11/23	47.8%	7/30	23.3%
Humerus[2]	3/7	42.9%	1/74	1.4%
Ulna[3]	N/A	N/A	8/48	16.7%
Radius[3]	15/18	83.3%	4/40	10%
Femur[3]	9/15	60.0%	6/46	13.0%
Tibia[4]	15/17	88.2%	9/33	27.3%
Fibula[4]	4/5	80.0%	4/20	20.0%
Astragalus[4]	23/35	65.7%	16/50	32.0%
Calcaneum[3]	9/11	81.8%	1/28	3.6%
Total	89/131	67.9%	56/369	15.2%

observed, "one must somewhere draw a completely arbitrary line, representing a point in time, across some steadily evolving lineage and say 'Here one taxon ends and another begins.'" Until further study, I must concur with H.G. McDonald (pers. comm.) in referring the *Megalonyx* material from the Camelot site to *Megalonyx* cf. *M. wheatleyi.* However, in light of this sample, as well as other currently unpublished data, the taxonomy of *Megalonyx* during the Late Pleistocene needs to be reassessed.

It is hopeful that my current research into rates of evolution and ratios of change in various elements of Pleistocene *Megalonyx* (Fields, unpublished data) will shed new light on the issue of taxonomy. Rates and processes of evolution are important to species concepts as they relate to the fossil record (Bell et al. 2004, Rose and Bown 1986). Other such studies (see several authors in Martin and Barnosky 1993) have provided valuable insights into paleontological species. The Camelot *Megalonyx* will be an integral part of the study of evolutionary rates, and no doubt the specimens of other taxa from the Camelot Local Fauna will be equally important and illustrative of the late Irvingtonian in the southeastern United States.

Acknowledgements

First and foremost I would like to thank Jim Knight of the South Carolina State Museum for allowing me to survey the outstanding Camelot collections, as well as for being a gracious and enjoyable host in his collections work area. I also thank Greg McDonald for providing unpublished data and also for his guidance in the identification of fossil elements and his valuable insights into ground sloth morphology. I would also like to thank the following curators and collections managers for access to comparative specimens: N. Gilmore, Academy of Natural Sciences at Philadelphia; R.C. Hulbert, Florida Museum of Natural History; J. Meng, American Museum of Natural History; R. Purdy, Smithsonian National Museum of Natural History; A.E. Sanders, Charleston Museum. I also wish thank D.A. Croft, G.F. Engelmann, and one anonymous reviewer for their comments that improved the quality of the manuscript. Thanks also go to The Culture and Heritage Museums, Rock Hill, South Carolina, for funding and research support.

References Cited

Bargo, M.S., De Iuliis, G., and Vizcaíno, S.F. 2006. Hypsodonty in Pleistocene ground sloths. Acta Palaeontologica Polonica 51(1):53–61.

Bell, C.J., E.L. Lundelius, Jr., A.D. Barnosky, R.W. Graham, E.H. Lindsay, D.R. Ruez, Jr., H.A. Semken, Jr., S.D. Webb, and R.J. Zakrzewski. 2004. The Blancan, Irvingtonian, and Rancholabrean Mammals Ages. Pp. 232–314 *in* Woodburne, M.O. (Ed.). Late Cretaceous and Cenozoic Mammals of North America: Biostratigraphy and Geochronology. Columbia University Press, New York.

Benton, M. J., and P. N. Pearson. 2001. Speciation in the fossil record. Trends in Ecology and Evolution 16(7):405–411.

Bruner, E. 2004. Evolution, actuality and species concept: a need for a paleontological tool. Human Evolution 19(2):93–112.

Chatters, J.C., S.H. Hackenberger, and H.G. McDonald. 2004. A Jefferson's ground sloth (*Megalonyx jeffersonii*) from the terminal Pleistocene of Central Washington. Current Research in the Pleistocene 21:93–94.

Christiansen, P., and R.A. Fariña. 2003. Mass estimation of two fossil ground sloths (Mammalia, Xenarthra, Mylodontidae). Senckenbergiana biologica 83(1):95–101.

Cope, E.D. 1871. Preliminary report on the Vertebrata discovered in the Port Kennedy Bone Cave. Proceedings of the American Philosophical Society 12(86):73–102.

Damuth, J., and B.J. MacFadden. (Eds). 1990. Body Size in Mammalian Paleobiology: Estimation and Biological Implications. Cambridge University Press, New York. 397pp.

DeQuieroz, K. 2005. Different species problems and their resolution. Bioessays 27:1263–1269.

Desmarest, M.A.G. 1822. Mammaloqie ou description des especes de mammiferes. Imprimeur Libbrarre rue des Poitevins 6. 555pp.

Fariña, R.A., S.F. Vizcaíno, and M.S. Bargo. 1998. Body mass estimations in Lujanian (late Pleistocene-early Holocene of South America) mammal megafauna. Mastozoología Neotropical 5(2):87–108.

Fields, S.E. 2009. Hypsodonty in Pleistocene *Megalonyx*: closing the diastema of data. Acta Palaeontologica Polonica 54(1):155–158.

Forey, P. L., R. A. Fortey, P. Kenrick, and A. B. Smith. 2004. Taxonomy and fossils: a critical appraisal. Philosophical Transactions of the Royal Society of London: Biological Sciences 359(1444):639–653.

Geisler, J.H, A.E. Sanders, and Zhe-Xiluo. 2005. A new protocetid whale (Cetacea, Archaeoceti) from the late middle Eocene of South Carolina. American Museum Novitates 3480: 1–65.

Gillette, D.D., H.G. McDonald, and M.C. Hayden. 1999. The first record of Jefferson's ground sloth, *Megalonyx jeffersonii*, in Utah (Pleistocene, Rancholabrean Land Mammal Age) *in* Gillette, D. D. (Ed.), Vertebrate Paleontology in Utah. Utah Geological Survey Miscellaneous Publication 99-1:509–521.

Gingerich, P. D. 1985. Species in the fossil record: concepts, trends, and transitions. Paleobiology 11(1):27–41.

Grine, F.E., W.L. Jugers, P.V. Tobias, and O.M. Pearson. 1995. Fossil *Homo* femur from Berg Aukas, northern Namibia. American Journal of Physical Anthropology 26:67–78.

Hirschfeld, S.E., and S.D. Webb. 1968. Plio-Pleistocene megalonychid sloths of North America. Bulletin of the Florida State Museum 12(5):213–296.

Hulbert, R.C. 2001. Mammalia 2: Xenarthrans. Pp. 175–187 *in* R.C. Hulbert (Ed.). The Fossil Vertebrates of Florida. University of Florida Press, Gainesville.

Jefferson, T. 1799. A memoir on the discovery of certain bones of a quadruped of the clawed kind in the western parts of Virginia. Transactions of the American Philosophical Society 4:246–260.

Kohn, M.J., M.P. McKay, and J.L. Knight. 2005. Dining in the Pleistocene: who's on the menu? *Geology* 33(8):649–652.

Kurtén, B., and E. Anderson. 1980. Pleistocene Mammals of North America. Columbia University Press, New York. 442pp.

Martin, R.A. 1974. Fossil mammals from the Coleman IIA Fauna, Sumter County *in* Webb, S.D. (Ed.). Pleistocene Mammals of Florida. The University of Florida Press, Gainesville.

Martin, R.A. and A.D. Barnosky. (Eds.). 1993. Morphological Change in Quaternary Mammals of North America. Cambridge University Press, New York. 415pp.

McDonald, H.G. 1977. Description of the osteology of the extinct gravigrade edentate, *Megalonyx*, with observations on its ontogeny, phylogeny, and functional anatomy. Unpublished Master's Thesis. Department of Zoology, University of Florida, Gainesville.

McDonald, H.G., and D. Anderson. 1983. A well preserved ground sloth (*Megalonyx*) cranium from Turin, Monona County, Iowa. Proceedings of the Iowa Academy Science 90(4):134–140.

McDonald, H.G., and C.E. Ray. 1990. The ground sloth, *Megalonyx*, from the North Atlantic Continental Shelf. Proceedings of the Biological Society of Washington 103(1):1–5.

McDonald, H.G. 1995. Gravigrade xenarthrans from the middle Pleistocene Leisey Shell Pit 1A, Hillsborough County, Florida. Bulletin of the Florida Museum of Natural History, Biological Sciences 37 Pt. II(11):345–373.

McDonald, H.G., C.R. Harrington and G. De Iuliis. 2000. The ground sloth, *Megalonyx*, from Pleistocene deposits of the Old Crow Basin, Yukon, Canada. Arctic 53(3):213–220.

McDonald, H.G., W.E. Miller, and T.H. Morris 2001. Taphonomy and significance of Jefferson's ground sloth (Xenarthra: Megalonychidae) from Utah. Western North American Naturalist 61(1):64–77.

McDonald, H.G. 2003. Sloth remains from North American caves and associated karst features *in* Schubert, B.W., J.I. Mead, and R.W. Graham. Ice Age Cave Faunas of North America. Indiana University Press, Indianapolis.

McDonald, H.G. 2005. Paleoecology of extinct xenarthrans and the Great American Biotic Interchange. Bulletin of the Florida Museum of Natural History. 45(4):313–333.

McHenry, H.M. 1992. Body size and proportions in early Homonids. American Journal of Physical Anthropology 87:407–431.

Mills, R.S. 1975. A ground sloth, *Megalonyx*, from a Pleistocene site in Darke Co., Ohio. Ohio Journal of Science 75:147–155.

Rose, K.D., and T.M. Bown. 1986. Gradual evolution and species discrimination in the fossil record. Pp. 119–130 *in* K.M. Flanagan and J.A. Liilegraven (Eds.). Vertebrates, Phylogeny, and Philosophy. Contributions to Geology, University of Wyoming, Special Paper 3.

Roth, V.L. 1990. Insular dwarf elephants: a case study in body mass estimation and ecological inference *in* Damuth, J., and B.J. MacFadden. (Eds.). Body Size in Mammalian and Paleobiology: Estimation and Biological Implications. Cambridge University Press, UK.

Ruff, C.B., W.W. Scott, and A.Y-C. Liu. 1991. Articular and diaphyseal remodeling of the proximal femur with changes in body mass in adults. American Journal of Physical Anthropology 86:397–413.

Simpson, G.G. 1928. Pleistocene mammals from a cave in Citrus County, Florida. American Museum Novitates 328:1–16.

Simpson, G.G. 1961. Principles of Animal Taxonomy. New York, Columbia University Press. 247pp.

Stock, C. 1942. A ground sloth in Alaska. Science 95(2474):552–553.

Von den Driesch, A. 1976. A Guide to the Measurement of Animal Bones from Archeological Sites. Peabody Museum Bulletin I. Peabody Museum, Cambridge, Massachusetts.

Woodburne, M.O. (Ed.). 2004. Late Cretaceous and Cenozoic Mammals of North America: Biostratigraphy and Geochronology. Columbia University Press, New York. 391pp.

Appendices

APPENDIX 1. Measurements of *Megalonyx* caniniforms from various localities. These data, along with those of *Megalonyx* caniniforms from Camelot L.F. (see Table 4) are plotted in Figure1. UL = upper left; UR = upper right; LL = lower left; LR = lower right.

Catalog #	Element	Locality	Length	Width	Comments
Megalonyx leptostomus					
UF 216900	caniniform-UL	Inglis 1A, Citrus Co., Florida	21.8	14.7	alveolar measurements
UF 216900	caniniform-UR	Inglis 1A, Citrus Co., Florida	24.4	15.3	alveolar measurements
UF 23587	caniniform-lower	Inglis 1A, Citrus Co., Florida	23.5	12.0	McDonald (unpublished data)
UF 23564	caniniform-LL	Haile 16A, Florida	28.6	14.8	
UF 23565	caniniform-LL	Haile 16A, Florida	29.5	14.1	
UF 23566	caniniform-UR	Haile 16A, Florida	28.2	17.5	
UF 20938	caniniform-LL	Inglis 1A, Citrus Co., Florida	24.2	10.5	
UF 21342	caniniform-LL	Inglis 1A, Citrus Co., Florida	22.6	11.8	
UF 23562	caniniform-LL	Inglis 1A, Citrus Co., Florida	20.9	10.7	
UF 20938	caniniform-LR	Inglis 1A, Citrus Co., Florida	24.7	10.6	
UF 223806	caniniform-LL	Haile 7G, Florida	24.5	12.5	in skull
UF 223806	caniniform-LR	Haile 7G, Florida	24.5	12.9	in skull
UF 223806	caniniform-UL	Haile 7G, Florida	23.6	14.4	in skull
UF 223806	caniniform-UR	Haile 7G, Florida	25.7	13.9	in skull
UF 223305	caniniform-LL	Haile 7G, Florida	29.1	13.2	
UF 223805	caniniform-upper	Haile 7G, Florida	30.6	15.6	McDonald (unpublished data)
UF ML-12	caniniform-LR	Haile 7G, Florida	23.9	14.2	McDonald (unpublished data)
UF 94624	caniniform-upper	MacAsphalt Shell Pit, Florida	18.1	13.2	
Megalonyx wheatleyi					
ANSP 19414	caniniform-LR	Port Kennedy Cave, Pennsylvania	·32.2	16.9	
ANSP 19415	caniniform-LR	Port Kennedy Cave, Pennsylvania	37.4	19.0	

Specimen	Tooth	Locality		
ANSP 19413	caniniform-LL	Port Kennedy Cave, Pennsylvania	35.3	15.4
ANSP 19416	caniniform-LL	Port Kennedy Cave, Pennsylvania	30.6	14.8
ANSP 19416	caniniform-LR	Port Kennedy Cave, Pennsylvania	32.3	15.0
ANSP 214	caniniform-LL	Port Kennedy Cave, Pennsylvania	34.5	16.1
ANSP 227	caniniform-LL	Port Kennedy Cave, Pennsylvania	30.6	15.1
ANSP 214	caniniform-LR	Port Kennedy Cave, Pennsylvania	34.1	16.8
ANSP 227	caniniform-LR	Port Kennedy Cave, Pennsylvania	29.2	13.7
SP 218	caniniform-LR	Port Kennedy Cave, Pennsylvania	36.4	18.2
USNM 11633	caniniform-LL	Port Kennedy Cave, Pennsylvania	30.4	16.1
USNM 11633	caniniform-LR	Port Kennedy Cave, Pennsylvania	30.3	16.9
ANSP 182	caniniform-UL	Port Kennedy Cave, Pennsylvania	33.7	19.4
ANSP 19430	caniniform-UL	Port Kennedy Cave, Pennsylvania	30.4	16.3
ANSP 19445	caniniform-UL	Port Kennedy Cave, Pennsylvania	31.0	17.8
ANSP 207	caniniform-UL	Port Kennedy Cave, Pennsylvania	33.3	17.9
ANSP 19226	caniniform-UR	Port Kennedy Cave, Pennsylvania	33.3	17.7
ANSP 19253	caniniform-UR	Port Kennedy Cave, Pennsylvania	34.5	19.3
ANSP 19258	caniniform-UR	Port Kennedy Cave, Pennsylvania	30.8	16.2
ANSP 19315	caniniform-UR	Port Kennedy Cave, Pennsylvania	30.0	16.5
ANSP 19318	caniniform-UR	Port Kennedy Cave, Pennsylvania	32.2	18.7
ANSP 19320	caniniform-UR	Port Kennedy Cave, Pennsylvania	31.7	18.0
ANSP 19326	caniniform-UR	Port Kennedy Cave, Pennsylvania	31.6	18.1
ANSP 19446	caniniform-UR	Port Kennedy Cave, Pennsylvania	33.1	16.0
ANSP 19449	caniniform-UR	Port Kennedy Cave, Pennsylvania	31.0	18.3
ANSP 203	caniniform-UR	Port Kennedy Cave, Pennsylvania	30.4	15.5
ANSP 19415	caniniform-UR	Port Kennedy Cave, Pennsylvania	36.1	17.7
ANSP 19414	caniniform-UL	Port Kennedy Cave, Pennsylvania	28.6	13.6
ANSP 19416	caniniform-UL	Port Kennedy Cave, Pennsylvania	35.1	18.3
ANSP 19413	caniniform-UL	Port Kennedy Cave, Pennsylvania	29.6	17.8
ANSP 19413	caniniform-UR	Port Kennedy Cave, Pennsylvania	34.8	18.0
ANSP 214	caniniform-UL	Port Kennedy Cave, Pennsylvania	33.6	18.6
ANSP 227	caniniform-UL	Port Kennedy Cave, Pennsylvania	30.5	17.3

Megalonyx wheatleyi (continued)

Catalog #	Element	Locality	Length	Width	Comments
ANSP 214	caniniform-UR	Port Kennedy Cave, Pennsylvania	32.1	18.8	
ANSP 227	caniniform-UR	Port Kennedy Cave, Pennsylvania	29.8	16.9	
ANSP 218	caniniform-UR	Port Kennedy Cave, Pennsylvania	30.0	15.4	
ANANSP 180	caniniform-LR	Port Kennedy Cave, Pennsylvania	30.8	17.1	
ANSP 180	caniniform-UR	Port Kennedy Cave, Pennsylvania	30.0	18.0	
ANSP 180	caniniform-LL	Port Kennedy Cave, Pennsylvania	31.6	17.2	
ANSP 15562	caniniform-LL	Port Kennedy Cave, Pennsylvania	29.4	14.7	
ANSP 19222	caniniform-LL	Port Kennedy Cave, Pennsylvania	32.8	18.0	
ANSP 19223	caniniform-LL	Port Kennedy Cave, Pennsylvania	33.9	17.6	
ANSP 19224	caniniform-LL	Port Kennedy Cave, Pennsylvania	32.0	15.8	
ANSP 19227	caniniform-LL	Port Kennedy Cave, Pennsylvania	32.8	15.4	
ANSP 19230	caniniform-LL	Port Kennedy Cave, Pennsylvania	26.9	15.4	
ANSP 19234	caniniform-LL	Port Kennedy Cave, Pennsylvania	33.8	15.3	
ANSP 19241	caniniform-LL	Port Kennedy Cave, Pennsylvania	33.9	18.4	
ANSP 19242	caniniform-LL	Port Kennedy Cave, Pennsylvania	29.8	13.0	
ANSP 19263	caniniform-LL	Port Kennedy Cave, Pennsylvania	26.8	16.0	
ANSP 19272	caniniform-LL	Port Kennedy Cave, Pennsylvania	29.5	15.2	
ANSP 19283	caniniform-LL	Port Kennedy Cave, Pennsylvania	32.1	16.3	
ANSP 19292	caniniform-LL	Port Kennedy Cave, Pennsylvania	35.1	16.2	
ANSP 19311	caniniform-LL	Port Kennedy Cave, Pennsylvania	37.9	15.3	
ANSP 19331	caniniform-LL	Port Kennedy Cave, Pennsylvania	26.1	11.5	
ANSP 19448	caniniform-LL	Port Kennedy Cave, Pennsylvania	32.5	16.0	
ANSP 19450	caniniform-LL	Port Kennedy Cave, Pennsylvania	34.0	15.9	
ANSP 200	caniniform-LL	Port Kennedy Cave, Pennsylvania	30.6	15.9	
ANSP 204	caniniform-LL	Port Kennedy Cave, Pennsylvania	29.9	14.2	
ANSP 205	caniniform-LL	Port Kennedy Cave, Pennsylvania	32.0	15.8	
ANSP 19271	caniniform-LR	Port Kennedy Cave, Pennsylvania	32.1	14.8	

Specimen	Position	Locality			Notes
ANSP 19310	caniniform-LR	Port Kennedy Cave, Pennsylvania	35.2	16.3	
ANSP 19314	caniniform-LR	Port Kennedy Cave, Pennsylvania	32.3	17.5	
ANSP 19322	caniniform-LR	Port Kennedy Cave, Pennsylvania	29.8	14.8	
ANSP 19329	caniniform-LR	Port Kennedy Cave, Pennsylvania	31.1	16.9	
ANSP 19447	caniniform-LR	Port Kennedy Cave, Pennsylvania	32.4	14.6	
ANSP 19225	caniniform-UL	Port Kennedy Cave, Pennsylvania	29.8	19.1	
ANSP 19264	caniniform-UL	Port Kennedy Cave, Pennsylvania	32.6	16.7	
ANSP 19274	caniniform-UL	Port Kennedy Cave, Pennsylvania	32.2	18.9	
ANSP 19295	caniniform-UL	Port Kennedy Cave, Pennsylvania	32.9	19.0	
ANSP 19296	caniniform-UL	Port Kennedy Cave, Pennsylvania	34.4	18.5	
ANSP 19330	caniniform-UL	Port Kennedy Cave, Pennsylvania	29.3	18.0	
UF 211058	caniniform-upper	Tri-Britton site, Hendry Co., Florida	30.6	16.0	McDonald (unpublished data)
UF 209395	caniniform-LR	Tri-Britton site, Hendry Co., Florida	36.4	16.1	
UF 21108	caniniform-UR	Tri-Britton site, Hendry Co., Florida	30.5	16.0	
F:AM 99187	caniniform-LL	McLeod Limerock Mine, Florida	30.9	16.4	in mandible
F:AM 99187	caniniform-LR	McLeod Limerock Mine, Florida	30.9	15.5	in mandible
F:AM 99190	caniniform-LR	McLeod Limerock Mine, Florida	34.2	15.6	in mandible
AM 99192	caniniform-lower	McLeod Limerock Mine, Florida	34.2	16.0	McDonald (unpublished data)
AM-103-1986	caniniform-lower	McLeod Limerock Mine, Florida	35.0	15.4	McDonald (unpublished data)
AM 102-1927	caniniform-LL	McLeod Limerock Mine, Florida	29.2	13.5	McDonald (unpublished data)
AM 102-1927	caniniform-LL	McLeod Limerock Mine, Florida	30.7	14.1	McDonald (unpublished data)
AM 102-1927	caniniform-LL	McLeod Limerock Mine, Florida	34.6	16.4	McDonald (unpublished data)
AM 104-1986	caniniform-LL	McLeod Limerock Mine, Florida	36.8	15.3	McDonald (unpublished data)
AM 102-1927	caniniform-LR	McLeod Limerock Mine, Florida	29.2	13.5	McDonald (unpublished data)
AM 102-1927	caniniform-LR	McLeod Limerock Mine, Florida	33.4	13.9	McDonald (unpublished data)
AM 102-1927	caniniform-upper	McLeod Limerock Mine, Florida	33.3	17.0	McDonald (unpublished data)
AM 102-1927	caniniform-upper	McLeod Limerock Mine, Florida	31.9	17.7	McDonald (unpublished data)
AM 75-1312	caniniform-upper	McLeod Limerock Mine, Florida	27.3	16.7	McDonald (unpublished data)
AM 91-1656	caniniform-upper	McLeod Limerock Mine, Florida	30.6	15.9	McDonald (unpublished data)
AM 99186	caniniform-upper	McLeod Limerock Mine, Florida	31.9	17.5	McDonald (unpublished data)
AM no #	caniniform-upper	McLeod Limerock Mine, Florida	29.8	17.6	McDonald (unpublished data)

Megalonyx jeffersoni

Catalog #	Element	Locality	Length	Width	Comments
USNM 424554	caniniform	Edisto Island, South Carolina	32.3	13.5	
USNM 424555	caniniform	Edisto Island, South Carolina	36.1	17.1	
ANSP 12536	caniniform-LR	Georgia	37.2	22.0	cast
ANSP 12536	caniniform-UL	Georgia	38.5	19.3	cast
ANSP 12536	caniniform-UR	Georgia	38.8	22.0	cast
ChM PV 5850	caniniform-LR	Coosaw River, South Carolina	34.6	18.6	McDonald (unpublished data)
SCSM 77.8.4	caniniform-LL	Cooper River, South Carolina	29.9	15.0	
SCSM 77.8.4	caniniform-LR	Cooper River, South Carolina	30.0	15.0	
USNM V175656	caniniform-UR	Pamlico Fm., Brevard Co., Florida	35.0	19.0	
ISM 492815	caniniform-UL	Lang Farm, Bureau Co., IL	39.3	17.5	
UF 213841	caniniform-UR	Aucilla River 1B, Florida	36.2	18.3	Alveolar measurements
UF 14832	caniniform-LR	Aucilla River, Florida	33.6	17.0	McDonald (unpublished data)
UF 23922	caniniform-lower	Santa Fe River 1, Florida	33.7	15.5	McDonald (unpublished data)
UF 23923	caniniform-lower	Santa Fe River 1, Florida	33.5	17.2	McDonald (unpublished data)
UF 10640-A	caniniform-upper	Santa Fe River 1, Florida	34.6	18.9	McDonald (unpublished data)
UF 23515	caniniform-lower	Waccassasa River, Florida	26.7	12.3	McDonald (unpublished data)
UF 23521	caniniform-lower	Waccassasa River, Florida	32.2	14.7	McDonald (unpublished data)
UF 16388	caniniform-upper	Waccassasa River, Florida	33.3	16.2	McDonald (unpublished data)
UF 23571	caniniform-lower	Warm Mineral Springs, Florida	30.8	15.6	McDonald (unpublished data)
UF no number	caniniform-lower	Warm Mineral Springs, Florida	32.5	15.5	McDonald (unpublished data)
UF no number	caniniform-LR	Warm Mineral Springs, Florida	30.0	15.4	McDonald (unpublished data)
UF 23568	caniniform-upper	Warm Mineral Springs, Florida	33.9	15.7	McDonald (unpublished data)
UF 92603	caniniform-LR	Aucilla River 2, Florida	38.5	17.2	McDonald (unpublished data)
UF 226237	caniniform-LL	Aucilla River 2C, Florida	33.6	17.2	McDonald (unpublished data)
UF 103492	caniniform-LL	Aucilla River 3J, Florida	36.0	16.1	McDonald (unpublished data)
UF 193956	caniniform-LL	Aucilla River 3J Florida	38.9	16.1	McDonald (unpublished data)
UF 153784	caniniform-UR	Aucilla River 3J Florida	38.6	17.3	McDonald (unpublished data)

Catalog No.	Tooth	Locality			Reference
UF 14906	caniniform-LL	Aucilla River, Jefferson Co., Florida	33.0	17.6	
ChM PV 5706	caniniform-UL	Edisto Island, South Carolina	36.7	17.8	
ChM PV 5708	caniniform-UR	Edisto Island, South Carolina	34.5	17.4	
UF V-4538	caniniform-UR	Ichetucknee River, Florida	34.1	16.5	
UF 36655	caniniform-LL	Oklawaha River 1, Florida	35.0	17.4	
UF 68179	caniniform-UL	Oklawaha River 1, Florida	35.6	18.7	
UF no #	caniniform-LL	Santa Fe River 1, Florida	34.6	15.1	
UF 10640	caniniform-LL	Santa Fe River 1, Florida	32.5	16.9	
UF SF2	caniniform-LL	Santa Fe River 2, Florida	32.7	17.0	
UF SF2	caniniform-UR	Santa Fe River 2, Florida	33.2	20.4	
UF 16388	caniniform-LR	Waccassasa River, Levy Co., Florida	32.8	16.1	
UF 23503	caniniform-upper	Waccassasa River, Levy Co., Florida	28.0	19.3	
UF 23504	caniniform-upper	Waccassasa River, Levy Co., Florida	34.9	16.8	
UF 16384	caniniform-UL	Waccassasa River, Levy Co., Florida	34.7	17.8	
UF 23507	caniniform-UR	Waccassasa River, Levy Co., Florida	33.7	14.8	
UF 23568	caniniform-LL	Warm Mineral Springs, Florida	33.1	16.3	
UF 23591	caniniform-LR	Warm Mineral Springs, Florida	30.0	14.8	
UF 23567	caniniform-UR	Warm Mineral Springs, Florida	39.4	18.7	
UF 16383	caniniform-lower	Waccassasa River, Florida	33.5	15.3	
CMN 24215	caniniform-LL	Old Crow Basin, Canada	34.4	14.6	McDonald (unpublished data)
CMN 31778	caniniform-UL	Old Crow Basin, Canada	34.4	20.0	McDonald et al. (2000)
CMN 48661	caniniform-UL	Old Crow Basin, Canada	31.5	15.4	McDonald et al. (2000)
USNM 10632	caniniform-upper	Kimmswick, MO	40.4	16.7	McDonald et al. (2000)
USNM 10632	caniniform-upper	Kimmswick, MO	40.2	17.9	
NWSM 002-28	caniniform-right	Grant County, Washington	32.0	16.0	Chatters et al. (2004)
UCMP 21429	caniniform-lower	Rancho La Brea, California	35.5	16.5	McDonald (1977)
USNM 1930	caniniform	Melbourne, Florida	38.9	16.8	
USNM 1925	caniniform-LL	Melbourne, Florida	33.5	16.3	
USNM 175655	caniniform-LR	Melbourne, Florida	42.5	17.3	
USNM 1925	caniniform-LR	Melbourne, Florida	40.1	16.4	
USNM 1925	caniniform-UL	Melbourne, Florida	42.3	20.6	
USNM 1930	caniniform-UL	Melbourne, Florida	41.1	18.8	
USNM 1925	caniniform-UR	Melbourne, Florida	44.9	22.1	

APPENDIX 2a. Measurements of the atlas vertebra   1. Greatest transverse width; 2. Greatest transverse diameter across the anterior of the articular facets; 3. Greatest dorsoventral width; 4. Anteroposterior length of middle of dorsal arch; 5. Greatest Anteroposterior length; 6. Anteroposterior length of middle of ventral arch; 7. Greatest transverse diameter of posterior articular facets

SCSM No.	1	2	3	4	5	6	7
2003.75.332	164.0	86.9	X	26.9	78.2	X	X
2004.1.111	147.3	91.3	79.6	31.3	75.0	23.5	89.8

APPENDIX 2b. Measurements of Vertebrae   1. Total height of vertebrae; 2. Length of centrum; 3. Width of centrum; 4. Height of centrum across posterior end; 5. Width of centrum across posterior end; 6. Length of neural spine from top of neural arch to top of spine; 7. Dorsoventral diameter of anterior end of neural canal; 8. Epiphyses fused? Y = Yes, N = No, Y/N = anterior epiphysis fused, posterior epiphysis unfused.

SCSM No.	Element	1	2	3	4	5	6	7	8
2003.75.324	vertebra, cervical	115.6	36.0	40.5	33.8	76.1	54.8	28.7	Y
2004.1.107	vertebra, cervical	103.3+	26.7	41.0	24.9	80.9	64.4+	27.4	N
2004.1.73	vertebra, cervical	X	35.3	51.7	32.0	88.3	X	26.3	Y
2004.1.99	vertebra, cervical	X	35.2	41.7	36.7	95.5	X	24.3	Y
2003.75.315	vertebra, anterior thoracic	X	30.2	43.5	32.7	X	X	X	N
2003.75.321	vertebra, anterior thoracic	X	35.0	47.7	32.1	X	X	X	Y/N
2003.75.323	vertebra, anterior thoracic	X	35.0	58.0	50.7	116.6	X	X	N
2003.75.325	vertebra, anterior thoracic	X	X	X	X	128.9	109.3	X	N
2003.75.326	vertebra, anterior thoracic	175.0	29.9	44.3	37.7	128.8	111.5	X	N
2003.75.437	vertebra, anterior thoracic	X	X	X	X	X	109.7	X	N
2003.75.438	vertebra, anterior thoracic	153.6	29.5	43.7	29.8	132.5	105.3	22.6	N
2004.1.74	vertebra, anterior thoracic	151.6	28.3	44.7	27.1	139.3	105.7	24.8	N

Specimen	Element								
2003.75.302	vertebra, posterior thoracic	187.1	43.2	68.9	59.4	122.7	113.4	X	N
2003.75.303	vertebra, posterior thoracic	175.8	43.9	70.8	62.1	128.8	75.9	41.3	Y/N
2003.75.304	vertebra, posterior thoracic	X	34.4	58.2	48.3	115.1	87.6	X	N
2003.75.305	vertebra, posterior thoracic	X	43.7	68.3	62.0	X	X	X	Y/N
2003.75.306	vertebra, posterior thoracic	X	37.9	59.4	54.3	X	X	X	N
2003.75.307	vertebra, posterior thoracic	X	39.3	71.9	60.8	X	X	X	N
2003.75.309	vertebra, posterior thoracic	X	53.2	84.3	63.7	X	X	X	Y
2003.75.311	vertebra, posterior thoracic	X	39.5	53.9	46.1	122.4	X	X	Y/N
2003.75.429	vertebra, posterior thoracic	X	37.1	66.9	51.2	X	X	X	N
2003.75.430	vertebra, posterior thoracic	X	38.8	67.6	58.3	X	X	X	N
2003.75.357	vertebra, posterior thoracic	X	36.0	60.0	53.9	X	X	X	N
2004.1.62	vertebra, posterior thoracic	160.4	38.8	66.4	54.5	135.0	62.8	40.5	N
2004.1.63	vertebra, posterior thoracic	X	36.6	67.3	51.8	117.9	X	40.0	N
2004.1.64	vertebra, posterior thoracic	157.4	38.3	68.2	56.5	133.2	65.7	40.0	N
2004.1.65	vertebra, posterior thoracic	190.0	48.6	74.5	59.1	135.6	77.7	43.4	Y
2004.1.66	vertebra, posterior thoracic	177.2	47.6	71.7	55.4	127.9	114.4	42.8	Y
2004.1.128	vertebra, posterior thoracic	X	X	X	X	116.4	100.7	X	X
2004.1.110	vertebra, posterior thoracic	153.6	38.7	67.9	56.0	132.4	80.1	45.4	N
2003.75.313	vertebra, thoracic, partial	167.8	39.2	67.1	59.4	130.2	94.1	41.9	N
2003.75.301	vertebra, lumbar	160.5	43.2	84.1	58.1	163.1	65.8	39.7	N
2003.75.427	vertebra, lumbar	X	38.4	71.6	48.9	X	X	X	N
2004.1.60	vertebra, lumbar	138.2	41.1	89.1	51.4	164.0	44.4	43.0	N
2003.75.316	vertebra, caudal	X	47.2	68.0	48.1	X	X	X	N
2003.75.314	vertebra, caudal	43.7	33.6	36.1	35.2	39.0	X	X	Y
2003.75.426	vertebra, caudal	X	40.5	60.4	53.0	X	X	X	Y/N
2003.75.428	vertebra, caudal	X	37.5	54.4	52.4	X	X	X	Y/N
2004.1.67	vertebra, caudal	63.8	32.2	55.1	50.5	59.3	5.8	8.6	N
2004.1.68	vertebra, caudal	45.3	38.6	44.8	38.4	45.3	X	X	Y
2004.1.69	vertebra, caudal	52.7	29.7	46.4	42.5	46.7	3.6	9.0	N
2004.1.70	vertebra, caudal	51.9	37.5	43.2	37.8	46.9	X	X	Y

SCSM No.	Element	1	2	3	4	5	6	7	8
2004.1.71	vertebra, caudal	59.6	40.1	47.3	45.5	54.3	4.8	8.0	Y
2004.1.101	vertebra, caudal	74.5	37.6	65.9	50.2	75.4	11.6	15.0	N
2003.75.317	vertebra, caudal, partial	X	44.2	54.7	55.1	X	X	X	Y
2003.75.318	vertebra, caudal, partial	X	39.4	57.3	52.4	X	X	X	Y/N
2003.75.319	vertebra, caudal, partial	X	49.9	68.7	57.8	X	X	X	Y
2003.75.320	vertebra, caudal, partial	~73.63	37.0	62.8	55.1	68.3+	X	~13.7	Y
2003.75.322	vertebra, caudal, partial	89.7	51.9	75.6	48.0	91.6	18.4	18.0	N
2004.1.72	vertebra, caudal, partial	45.4	~27	40.0	36.1	41.1	X	X	Z
2003.75.308	vertebra, centrum	X	42.8	94.0	53.9	X	X	X	Z
2003.75.310	vertebra, centrum	X	37.6	59.3	51.2	X	X	X	Z
2003.75.312	vertebra, centrum	X	52.5	82.6	63.7	X	X	X	Y
2004.1.61	vertebra, centrum	X	35.9	60.4	49.9	X	X	X	N

APPENDIX 3. Measurements of *Megalonyx* phalanges from Camelot L.F. of unknown position on the manus or pes. 1. Maximum proximodistal depth; 2. Maximum mediolateral width; 3. Maximum anteroposterior length; 4. Epiphyses fused?

Catalog #	Element	1	2	3	4	Comments
2003.75.338	phalanx-distal	110.3	28.0	39.2	Yes	
2003.75.339	phalanx-distal	125.4	25.0	51.3	No	epiphysis included
2003.75.339	phalanx-distal	118.9	20.3	45.6	No	epiphysis included
2003.75.342	phalanx-distal	133.1	24.7	46.1	No	epiphysis included
2003.75.343	phalanx-distal	126.6	33.7	56.9	Yes	
2003.75.344	phalanx-distal	117.5	26.2	44.8	Yes	
2003.75.345	phalanx-distal	124.8	26.6	50.9	Yes	
2003.75.346	phalanx-distal	93.5	26.0	53.7	No	prox. epiphysis missing
2003.75.347	phalanx-distal	117.7	32.2	51.6	Yes	
2003.75.406	phalanx-distal	162.0	34.4	59.6	Yes	
2003.75.408	phalanx-distal	123.0	30.3	40.8	Yes	
2004.1.123	phalanx-distal	84.9+	26.3	46.0	No	prox. epiphysis missing
2004.1.83	phalanx-distal	102+	19.2	44.3	No	prox. epiphysis missing
2004.1.85	phalanx-distal	96.5	20.3	41.5	No	articular surface missing
2003.75.445	phalanx-distal	35.4	8.9	26.9	No	epiphysis missing
2003.75.448	phalanx-distal	37.8	13.1	33.5	No	epiphysis missing
2003.75.375	phalanx-medial	74.1	34.8	46.4	Yes	
2004.1.104	phalanx-medial	51.8+	25.2	35.7	No	prox. epiphysis missing
2004.1.35	phalanx-medial	59.4+	33.5	39.6	No	prox. epiphysis missing
2004.1.52	phalanx-medial	82.8	31.3	41.7	Yes	
2004.1.53	phalanx-medial	60.6	29.0	38.5	No	
2004.1.55	phalanx-medial	59.1	32.6	39.4	No	
2004.1.58	phalanx-medial	66.1	30.9	38.6	Yes	
2003.75.349	phalanx-proximal	35.5	33.1	50.4	Yes	
2003.75.350	phalanx-proximal	30.4	33.5	45.8	Yes	
2004.1.116	phalanx-proximal	29.1	34.2	47.7	Yes	
2004.1.117	phalanx-proximal	31.8	35.3	46.1	Yes	
2003.75.379	phalanx-proximal	31.8	39.1	50.6	Yes	
2003.75.378	phalanx-proximal	29.9	39.2	51.6	Yes	
2003.75.381	phalanx-proximal	28.8	37.5	52.2	Yes	

Index

www.ingramcontent.com/pod-product-compliance
Lightning Source LLC
Chambersburg PA
CBHW050349110426
42812CB00008B/2420